Governing Disaster in Urban Environments

Governing Disaster in Urban Environments

Climate Change Preparation and Adaption after Hurricane Sandy

Julia Nevárez

LEXINGTON BOOKS
Lanham • Boulder • New York • London

Published by Lexington Books
An imprint of The Rowman & Littlefield Publishing Group, Inc.
4501 Forbes Boulevard, Suite 200, Lanham, Maryland 20706
www.rowman.com

Unit A, Whitacre Mews, 26-34 Stannary Street, London SE11 4AB

British Library Cataloguing in Publication Information Available

Library of Congress Cataloging-in-Publication Data

Names: Nevárez, Julia, author.
Title: Governing disaster in urban environments : climate change preparation and adaption after Hurricane Sandy / by Julia Nevárez.
Description: Lanham, Maryland : Lexington Books, 2018. | Includes bibliographical references and index.
Identifiers: LCCN 2018006097 (print) | LCCN 2018003879 (ebook) | ISBN 9781498543767 (electronic) | ISBN 9781498543750 (cloth) Subjects:
LCSH: Emergency management—United States—Planning. | Natural disasters—United States—Planning. | Assistance in emergencies—United States—Planning. | City planning—Environmental aspects—United States. | Hurricane Sandy, 2012.
Classification: LCC HV551.3 (print) | LCC HV551.3 .N478 2018 (ebook) | DDC 363.738/748—dc23
LC record available at https://lccn.loc.gov/2018006097

♾™

Contents

Acknowledgments

I would like to express my deep appreciation and gratitude to students, colleagues, friends, and family whose support and inspiration made this book possible. I am most appreciative of Facebook groups and friends who provided invaluable direct and indirect virtual support. This book is dedicated to the communities of solidarity that emerge after extreme climate events.

To Jorge and Reina.

Introduction

Hurricane Sandy and the Challenges of Governing Disaster

Both objects of study, the city and climate change require an interdisciplinary perspective. It is the intention of this book to provide an introduction to relevant discourses, issues, and questions that bear on the complexity of climate change in urban areas. The book seeks to frame questions and to provide openings for strategies and practices regarding preparedness and adaptation to climate change in urban environments after Superstorm Sandy's impact in 2012. Sandy was a shocking reality and one that seems to repeat itself after the 2017 hurricane season that broke records both in terms of quantity and intensity. Hurricanes Harvey, Irma, and Maria, like Sandy, made blatantly visible the precariousness of the infrastructure and planning flaws, ill placed psychological predispositions, and the uneven impact these systems have in areas already facing challenges in terms of social, economic, and environmental disadvantages. This book seeks to articulate a framework as a point of departure toward finding common ground, critical perspectives, and strategies that work considering the multi-scalar phenomenon that constitutes climate change in the city. As complex phenomena, both climate change and the urban condition require a multidisciplinary approach and the focus has been to bring to our attention what could be some of the most relevant perspectives on governing disaster. The perspectives entertained in this book could be considered somehow far reaching and to a certain extent ambitious, but multiple are the factors that impinge on the governance of disaster. At this historical juncture, however, pressing set of circumstances show palpable, visible, and immediate climate change impacts, adding a sense of urgency to consider many possible relevant bodies of knowledge, experience, and expertise. Research on climate change is often technical and discipline specific. Even though these do provide in-depth analysis, they lack a comprehensive and overarching robust approach that takes into consideration the social

Figure I.1 Midtown Manhattan Skyline, New York City, 2017. Photo by J. Nevárez.

component, especially. The different bodies of literature included in this book range from geography to sociology, planning, architecture, political science, and environmental sciences. The disciplinary frameworks selected and the literature examined in this book seek to account for the multi-scalar levels by which climate change as a phenomenon affects cities and to account for the degree of complexity that such an endeavor requires. In the attempt to provide a point of departure for scholarship on climate change where multi-scalar conceptualizations offer alternatives based on communities of solidarity, voluntary and involuntary omissions along with possible misinterpretations are the responsibility of the author.

Climate change poses many challenges in terms of recovery, preparedness, and adaptation from both extreme climate events and slow-onset emergencies. Issues of vulnerability and risk lay at the forefront of concerns to address climate change in terms of how they are perceived (Abid et al. 2016; Kettle and Dow 2016); as abrupt and nonlinear (Briggs 2017); in the methodology of mapping for disaster preparedness (Busby 2017); conceptualized within the reflexive risk ethos of modernity (Ekberg 2007), and complex and systemic risks (Mourad and Luers 2017). Environmental security calls for broad understandings about populations and ecosystems, political boundaries, and how climate change impacts management, among many other emerging governance concerns that would likely intensify with climate change impacts (Barnett 2003; Barnett and Adger 2007; Buzan et al. 1998; Dabelko 2006; Dalby 2013; Environmental Change and Security Program 2016; O'Callahan 2012; Vivekananda and Beat 2017; Werrell and Femia 2017). Adaptation is at the forefront of concerns about climate change impacts (The Adaptation Fund Board Secretariat 2016; Adger et al. 2005; Berkes and Jolly 2001; Biesbroek et al. 2010; Gonzalez et al. 2016; Hughes and Sarzynski 2015; Tompkins and Adler 2004; Simonet and Fatoric 2016). A resilient culture where communication, strategies, and practices of resiliency are constantly expressed, shared, and explored would provide a degree of preparedness necessary for extreme climate change events (Adger 2000; Aldrich 2012; Bernes and Folke 1998; Gunderson and Holling 2010; Klein et al. 2004; Saloojee and DeBritto 2014).

Many have undertaken the opportunity to critically examine and produce alternative conceptualizations to the relationship between humans and nature. Industrialization and development are then seen as the conditions that produced environmental problems, characteristically urban. More recent conceptualizations originate in notions such as the "Anthropocene" to press emphasis on the seemingly geological force-like by which humans have impacted the earth's ecology, and in this way open the discourse to make stark and necessary connections between geological and social sciences (Adeney 2014; Bennet 2010; Chakrabarty 2012; Crutzen and Stoermer 2000; Dalby 2007; Haraway 2016; Johnson and Morehouse 2014; McKensie 2015; Mirzoef 2014; Moore 2016; Palsson and Swanson 2016; Purdy 2015; Thomas 2014; Yusoff 2015). Among these, the relationship between organic and inorganic matter, nature and humans, and how populations under neoliberalism within a state of insecurity are managed and administered as a form of governmentality are core conceptual concerns that guide the arguments in this book largely drawing from the experience of Hurricane Sandy in New York City.

The Anthropocene has generated a reconsideration of alternative discourses where humans and nature are to respond to a different kind of relationship than the one inherently limited within industrialization and capitalism where nature is considered a resource to be exploited. Moreover, the paradox of security and insecurity in times of scarcity, vulnerability, and risk constitutes a resilient subject managed to absorb change in a constant state of experimentation producing, among other things, an entrepreneurial subject within an entrepreneurial state and city. Forms of governance in post-Fordist societies within neoliberal frameworks and the constitution of the subject in a resilient care of the self are fitting to the multi-scalar processes of globalization and the challenges posed by the transnational and trans-local character of climate change relationships. Betsill and Bulkeley (2006) and Corfee-Morlot et al. (2009) point at the multi-scalar level of governance emergent from neoliberal practices and addressed in the literature to convey forms of meta-governance (Davies 2014) involved in the multilevel governance (Homsy and Warner 2015; Kern and Harriet 2009) in regimes of complex political scales (Koohane and Victor 2011). But it is in the context of neoliberalism's political project that the pervasiveness of market deregulation and privatization many argue with more or less enthusiasm, lay the limits to and possibilities for alternative forms of governance (Barnet 2003; Barnet et al. 1996; Bogaert 2013; Brady 2016; Harvey 2007; Brockling 2016; Brenner 2004; Brenner and Theodore 2002, 2005; England and Ward 2007; Ferguson and Gupta 2002; Jessop 2007; Larner 2003). Issues of social justice emerge as an important consideration to the administration and management of populations (Smith et al. 2008; Springer 2010, 2015; Swyngedouw 2007; Swyngedouw et al. 2002; Zamora 2016) where cities become important nodes (Peck 2001, 2002; Peck and Tickell 2002; Peck et al. 2009). Foucault's notion of governmentality is a useful analytical tool to understand how neoliberal state governance is justified (Foucault 1991, 1994, 2008; Lemke 2002; Vitermark 2005). Similar to neoliberal governance, climate change also involves a multi-scalar approach. Both neoliberal governance and climate change management function on a trial and error, experimental base with entrepreneurial forms of policy development localized in cities (Brockling 2016; Broto and Bulkeley 2013; Davis 2010; Kern and Bulkeley 2009; Kemmerzell 2016).

Hurricane Sandy is one of the most recent extreme climate events in the Eastern United States triggering a serious discussion on preparedness for climate change. Sandy and the conditions under which rebuilding takes place are introducing new epistemologies and necessary changes to assumed ways of thinking about natural, built, and social environments. Many possible applications and approaches to climate change preparation and adaptation have emerged. While most of the literature makes reference to infrastructural applications, many argue that nonstructural, social, and community alternatives deserve closer attention (Blachman-Biatch et al. 2013; Dimmer and Lindberg 2014; Gonzalez et al. 2016; Gupta et al. 2016; Jaleel 2012).

The Anthropocene identifies the impact of urban industrial society in the deforestation, overfishing, ozone depletion, mangrove conversion, and agricultural expansion from which the richest countries generate a significant percentage of the environmental degradation that contributes to climate change. This opens the discourse on social and climate change to further considerations about the relationship between humans, nonhumans, and organic matter. From the perspective of the Anthropocene, industrial human activity and modernity's ideologically myopic vision, short-term goals and focus, produced landscapes destructive to the environment with such a force that a geological age has been designated to articulate this rift and to highlight the need for an epistemological shift. The Anthropocene can be considered a critical space to produce a stance from which to generate new forms of understanding about humanity's relationship with nature and advocate for an earth-friendly sociality based on the questioning of human interference with nature's cycles through interventions that diminish human and natural diversity on planet earth. Industrialization, rapid development of agriculture, and urban civilization along with energy use based on fossil fuels all contribute to the tipping point in 2010 of 385 parts per million (ppm) of carbon in the atmosphere that is destroying the planet's biota. The urbanization of humanity and subsequent destruction of the natural environment makes of cities the problem but also the solution. This paradox sheds light into the possibility of considering the city as a transformative site with the ability to implement a sustainable and greener urbanism alongside an ethos of human solidarity with the environment as a form of global commons.

Strategies and policies in these challenging times are required to address climate change conditions within the also challenging practices of a neoliberal state disengaged from economic support for commonly shared risks and vulnerabilities. The administration and management of populations during times of crises produces a precariousness that clogs the system in the delivery of aid to communities affected by extreme climate events like Hurricane Sandy. Different to Katrina's response and despite Hurricane Sandy's immediate overwhelmed infrastructure and lack of preparation to a somehow unexpected climatic event of Sandy's magnitude, attention and resources were directed to New York global city. However, economically disadvantaged neighborhoods and communities were doubly affected by Hurricane Sandy's impact much in tune with the generalized pattern of income polarization generated by globalization throughout the planet. Those who most needed government aid received a limiting militarized approach from Homeland Security that showed organizational shortcomings when working alongside communities who faced the multiple challenges posed by Superstorm Sandy. The government's response did not match the magnitude of the event. This experience highlighted a complex set of issues and factors that would likely affect future responses to climate change, namely, an understanding of political and economic contextual factors for the management of populations in constant man-made and natural crises modes. The models of administration and management currently in place based on an ideology of resilience to absorb and bounce back confounded with economic and climatic upheavals challenge present social structures and physical infrastructures.

Traditional governmental organizational approaches were flawed in the challenge to access and distribute aid to affected communities impacted by Sandy. However, assemblages of community support that groups like Occupy Sandy coordinated in multiple relief hubs throughout the affected areas in the city became indispensable to fill in the gaps and often lead recovery strategies and aid distribution, especially to neighborhoods with already limited resources. Occupy Sandy's organizational model based on solidarity, basic human connection, and intelligence coordinated rescue efforts efficiently and effectively based on a horizontal model that skillfully used social media and provided a bottom-up approach. In this regard, Occupy Sandy by coordinating with other individuals and organizations was closer to the "Whole Community" approach that the Federal Emergency Management Agency's (FEMA) aims to develop but was challenged to apply during Sandy. FEMA's largely hierarchical top-down command and control organizational approach was quickly overwhelmed by the storm's impact and the magnitude of the implementation challenge with many, most vulnerable community sectors left unattended.

Still an ongoing process, Sandy has opened a space for dialogue on social and spatial reconsiderations for our urban mode of living in today's changing climate. Hurricane Katrina was the first extreme climate change event that led to a serious analysis of climate change preparation. Katrina also made visible the length of time an event of such magnitude lingers in the social, cultural, economic, and spatial constitution of the areas affected. These climate change impacts have lasting consequences. Housing stock destruction, population shifts, and increase in civic participation in a post-Katrina New Orleans are similar changes to those found after Hurricane Sandy in New York City and New Jersey.

A discussion about climate change will offer the context to discuss the emergence of the Anthropocene conceptual alternatives in chapter 1, "Climate Change, Katrina, and Sandy." The chapter will also address environmental security as an emergent field that considers the risk and vulnerabilities posed by climate change and the organizational structure in the governance of aid embedded in the governmental structure of the United States regarding the delivery of aid after an extreme climate event. Sustainability at times used as a public relations strategy for governments and companies to justify policies and budgets can be considered part of the environmental security framework. Chapter 1 will conclude with an overall account of Hurricane Sandy's impact on the Eastern Seaboard of the United States, especially New York City.

Chapter 2, "Neoliberalism: Globalization and the Urban Paradox in Climate Change Governance," will examine the role of the neoliberal state at the international, national, urban, and subject levels of governance. This multi-scalar framework constitutes the scope of extreme climate governance within neoliberal entrepreneurial policy practices. The emergence of International Climate Agreements as a supra-structure to legitimize, regulate, and administer climate change preparation and adaptation policies is a form of climate governance at the international level that will also be examined in chapter 2. The social, economic, and environmental issues related to extreme climate place cities and environmental sustainability into a preeminent focus. Extreme climate events impact urban life, altering everyday cycles and affecting the productive capacity of its dwellers in significant ways as exemplified by Hurricane Sandy. The city can then be considered a paradox of problematic practices. The city emerged from industrialization and the excesses of modernity's ideology of progress and development. But it is largely in cities where policies to extreme climate tend to be crafted and implemented. The city is the cradle where experimentation with innovative practices to prepare, adapt, and respond to extreme climate events usually takes place. The "urban" is then the node where levels of governance and climate change simultaneously intersect. The ecological city constituted by the physical infrastructure, organic and inorganic matter, as well as humans is a fresh reminder of the possibilities that the urban offers in terms of awareness, preparation, and adaptation to climate change impacts.

Chapter 3, "Post-Sandy Disaster and Climate Governance: Resiliency in Adaptation and Mitigation," will discuss the discourse on resiliency that emerged in New York City after Hurricane Sandy and has largely been associated with preparedness and adaptation. An event of Sandy's magnitude produced an event of similar magnitude on reflexivity. Addressing immanent change—climatic and social—has triggered a renewed consideration on strategic development that is flexible and redundant. Some of the new terminology that has emerged after Sandy is resiliency and redundancy. Both are terms widely applied within the discourse on preparedness and recovery not only in the United States but also in cities around the world. This emerging discourse includes the possibility to constantly rethink and evaluate options that might otherwise be considered as failures when thinking about multiple systems that require attention before and after a climate event of Sandy's scale. The conditions under which rebuilding takes place are based on innovative policies where resiliency is considered a form of bouncing back from an extreme event and absorbing change a form of adaptation. Reinvention and experimentation are part of the entrepreneurial character of both neoliberal governmentality and the development of climate change preparation and adaptation policies. The rationality implemented to manage and administer populations rests on a care of the self to produce a subject that is a resilient individual also. Climate change policies are also based on an experimental mode. The process of seeking answers where many possible applications and approaches have emerged emphasizes the multi-scalar and trans-local component on knowledge transfer that informs discourses about preparedness, adaptation, and recovery.

In chapter 4, "Alternatives: Social and Physical Redundancies," social and spatial alternatives to address climate change impacts, recovery, preparedness, and adaptation are examined. Social redundancies at the community level refer to alternatives such as the one exemplified by Occupy Sandy. Their relief hub model emerged as a site to coordinate vital resources to assist and subsequently help rethink the important role of community. Their approach helped fill the gap left at the local level by more traditional command and control top-down governmental approaches such as that of FEMA and Homeland Security. The flexible and horizontal approach based in solidarity guided the Occupy Sandy model, which proved to be intelligent, skilled, efficient, and effective mostly in communities that generally lack resources. The need for cultures of resiliency at the community level that are able to co-create responses to extreme climate events promptly and efficiently by collaborating among different groups and organizations is part of an alternative social redundancy approach. Localized community responses in Japan after the tsunami will also be discussed as an example of community redundancy and resiliency. Chapter 4 will also examine varied but non-exhaustive national and international examples that highlight physical infrastructure redundancies in the form of flood-prone design using both built infrastructure and natural ecosystems.

Drawing from the literature and the experience of recuperation after Sandy in New York City, chapter 5, "Climate Change: Resilient and Redundant Social Policy," will offer social policy recommendations to develop resilient cultures based on effective communication, inclusiveness, participation, public and private collaborations, forward-looking redesign strategies by interdisciplinary and multidisciplinary cadres of professionals among other social policy considerations.

The challenges of addressing individual, social, cultural, and economic aspects involved in the multi-scalar impact of climate change will be examined in the concluding chapter, "Tackling Climate Change: The Epistemological Rift in Governing Disaster." The epistemological change that climate change poses requires a resilient and adaptive inclination and a ground-up approach to communal living and organization. The ontological aspect of development following a linear path in the cultural logic of modernity needs to be questioned along with the neoliberal model of market deregulation and privatization. The neoliberal model that awards individualism under constant economic, social, cultural, man-made, and environmental/climatic crises is insufficient and limiting. Five years after Sandy, survivors do not consider that we are prepared and mental health issues figure as one of their top priorities in terms of available support during and after an extreme climate change impact (Murray 2017). Climate change crises alter urban environments and city life. Cities are key nodes for climate change policy interventions. The political ecology of extreme climate impacts produces landscapes prone to privatization and displacement from reinvestment and redevelopment. Even though climate change policy interventions tend to prioritize on infrastructure, social and community, nonstructural components are key aspects for societies that thrive with extreme climate impacts as they will likely increase in quantity and intensity. These events highlight the underlying social and economic conditions of infrastructures and community, and the uneven geography of development along class and race indicators. The required adaptation to new ways of thinking and practices about the natural environment and cities emerges in an ecological approach constituted by relationships of collaboration between humans and nonhumans, organic and inorganic matter. The environment as a global commons offers a point of departure to establish relationships of solidarity among people and their biota aimed at protecting and enhancing such relationship on an equal footing in a globalized planet. The political ecologies of such commons are based on alternatives where communities of solidarity are strengthened before and after extreme climate events.

Chapter 1

Climate Change, Katrina, and Sandy

CLIMATE CHANGE

Debates about the facticity of climate change are challenged by the ample evidence and awareness about increased extreme climate impacts experienced around the world recently. Many accept these unusual patterns of increased climate intensity as examples of human intervention on ecosystems while others remain skeptic. According to Wuebbles et al. (2017) from the US Global Change Research Program, greenhouse gases emitted from human activities are the main cause of annual global average surface temperature increase in the last 115 years to denominate the warmest period of modern civilization. Changes in surface, atmospheric, and oceanic temperatures, melting glaciers, diminishing snow cover, shrinking sea ice, rising sea levels, ocean acidification, and atmospheric water vapor have been documented by many scientists and researchers (Wuebbles et al. 2017). Evidently, climate change sea level rise and intense warm temperatures around the globe tend to increase the extent and magnitude of coastal floods caused by hurricanes. The most impactful storm in many years, Hurricane Sandy has brought public attention to the climate hazards of the New York area specifically, and to climate change more generally. Many argue that Hurricanes Katrina and Sandy are examples of the threatening variability of the climate due to human impacts. Climate change has been acknowledged as part of our contemporary global experience. Climate overcomes territorial boundaries and presents challenges to the all-encompassing connections among peoples, places, and ecologies. This chapter will address general conceptualizations about climate change in order to understand the current state of climate change discourse.

Climate changes carry consequences to poor neighborhoods and countries unevenly and pose challenges to the world's most advanced societies

Figure 1.1 Impact of Hurricane Sandy in the Rockaways, New York City. 2012. Photo by J. Nevárez.

affecting international relations, migration patterns, and the state of democracy around the world. The Anthropocene and Capitalocene are considered distinct eras to indicate how human activities have altered the conditions of the planet's organic and nonorganic matter, its topography and weather, based on the industrialization of the economy. The acknowledgment of human impact consequences around the globe directs us to consider the need for an alternative paradigm to face the risks, vulnerability, and uncertainty of climate change. Under those circumstances, environmental security and the administration of aid will be addressed to understand the role of the government in the provision of aid and recovery. The military paradigm usually applied to extreme climate impacts is a form of governmentality that in the United States is administered under the Department of Homeland Security to manage populations and resources affected by both man-made and natural disasters. The organizational chain of command for the distribution of aid under extreme circumstances within the context of the United States will then be discussed. The chapter will continue with a careful examination of two of the most relevant and recent climatic events in the Eastern Seaboard of the United States: New Orleans's Hurricane Katrina and New York City's Hurricane Sandy.

Climate change involves a complex set of aspects. Population growth requires the use of energy sources that generate pollution by carbon dioxide (CO_2) emissions, greenhouse gases, and ozone layer disruption. According to Wuebbles et al. (2017), the global atmospheric CO_2 concentration exceeded 400 parts per million (ppm), a level that happened about 3 million years ago at a time when sea level and global average temperature were much higher than current figures. These circumstances are the leading causes of global warming and the sea level rise observed in the last century. Warmer oceans contribute to extreme climate events like hurricanes, and the melting of the ice caps is conducive to sea level rise. The overemphasis on fossil fuels for energy has been identified as one of the main causes of pollution and environmental degradation. Largely, the pollution of greenhouse gases in the atmosphere stems from the consumption of fossil fuels for industrial production and travel, and from new forms of land use and agriculture. Giddens (2011) identifies two different paradigms in regard to pollution: the old industrial paradigm based on fossil fuels and the "green paradigm" based on clean energy. In the old industrial paradigm, advocating for competitiveness and the lack of environmental regulatory restrictions or taxes in developing countries increased environmental problems. Developed countries tend to be the largest polluters by their higher levels of energy consumption in support of higher standards of living for a larger segment of their population. Increasingly so, and due to the shift in the economy to outsource manufacturing, developing nations including China and India also became large polluters. The developing world then faces similar challenges to those of developed countries in terms of climate change drivers. The detrimental effects to the environment and the intensification of extreme climate events are due to fossil fuels based energy production. Economic growth, resource scarcity, population increase, the expansion of low-carbon technologies, and the intensifying of regional inequalities are all conducive to the detriment of the environment. The wealth produced by fossil fuels for energy largely falls in the hands of local elites (Giddens 2011).

Human activity has been a determining contributor to the observed warming of the earth by introducing greenhouse gases in the atmosphere by the consumption of fossil fuels, industrial production, air traffic, shipping, urban-suburban development, and large-scale agriculture (Giddens 2011). The growth in world population predicted to reach 9.6 billion by 2050 will place enormous strain on global resources. For a sustainable future, cleaner forms of energy generation that do not involve fossil fuels and management of population growth are crucial factors in order to decrease carbon emissions and limit global warming and greenhouse gases. An unpredictable climate transition with progressively intensifying and generalized climate event impacts are currently observed not only from computer-generated models that measure risk, but from recent impacts, especially during the 2017 hurricane season. The warming of the oceans, the rising sea levels, and the melting of the glaciers and of the Arctic ice cap are considered to be directly influenced by human activities, accelerating climate change (Giddens 2011). The depletion of fossil fuel resources is led by a growing demand for energy production by drilling in remote places and at greater depths under the sea and waterbeds.

The acknowledgment of the need to integrate energy policy with the effort to control climate change has become a generalized concern, mostly in the developed world and increasingly so in the developing world. Energy consumption of oil and gas by the United States is the greatest. According to the US Energy Information Administration,[1] energy consumed in 2015 (measured in quad btu) equaled 79.2891 in fossil fuels, 8.337 in nuclear electric power, and 9.6746 in renewable energy for a total of 97.5282 btus. US energy produced 70.046816 btus in fossil fuels, 8.33768 btus in nuclear electric power, and 9.693623 btus in renewable energy for a total of 88.078118 btus. China and India are increasingly becoming stronger energy consumers, and specifically China buys energy, like Britain and the United States, from other countries such as Venezuela, Indonesia, Iraq, Oman, Yemen, and Sudan (Giddens 2011).

The United States is one of the largest polluters with approximately 4 percent of the world's population, 25 percent a consumption of global energy, and generates 20 percent of the world's carbon emissions (Giddens 2011). The role of new technologies is crucial in order to diminish carbon emissions into the atmosphere and governmental incentives to produce such energies are vital. Emphasis on the transformation into new technologies to produce energy rests on low-carbon technologies such as purified coal (carbon capture and sequestration, or CCS), wind power, tidal or wave power, biofuels, solar power, geothermal energy, smart electricity grids, geo-engineering technologies such as heat shields that would turn back a proportion of the sun's rays, and "scrubbers"—devices that would extract COs and other greenhouse gases out of the atmosphere. The economics of innovation is an effective area for government intervention while creating jobs in renewable technologies. At this point this transition is still costly.

The Brundtland Report in the early 1970s introduced the notion of sustainability in an effort to deter environmental problems and diminish the intense use of natural resources. Sustainability rests on the assumption of preserving natural ecosystems to promote human welfare by satisfying basic needs with consideration for future generations. Environmental sustainability focuses on the improvement of air, soil, and water as part of the ecological systems and to decrease pollution, food scarcity, and diseases. Sustainability also considers building on social and institutional capacities to cope with environmental hazards as a way to develop a sense of stewardship toward global public goods including the atmosphere as a global commons (Giddens 2001). The dynamic equilibrium supported by an environmental sustainability framework is based on limits to development. Consumption remains one of the main drivers of pollution and in advanced countries' relatively higher purchasing power, one of the main factors that contribute to pollution. The 2006 Sustainable Society Index (SSI) provides wider range of environmental measures that includes "resource depletion affecting wetlands, forests, farmland and non-renewable raw materials, higher levels of carbon emissions and other potential causes of environmental damage" (Giddens 2011, 65). Despite concerted efforts to overcome the dire situation regarding climate change, the number of catastrophes in the world has risen and the case has been made that a crucial component to the incidence of extreme climate is due to the rise of global temperatures. Hurricanes Katrina and Sandy have been very costly in addition to the most recent extreme climate generated by Harvey, Irma, and Maria. The financial damages of such catastrophes involve extreme losses. Currently the cost of recovery from extreme climate impacts is jeopardizing the budget capacity of the US Federal Emergency Management Agency in terms of recovery and insurance claims for those areas that have an insurance framework from where to grant coverage.

CLIMATE CHANGE: RISKS AND VULNERABILITIES

The risks posed by climate change events and those that increase vulnerability center around food insecurity, water availability, and livelihood disruption. Natural disasters provide the opportunity to examine more acutely the taken-for-granted assumptions regarding ways of thinking, modes of existence, and economic arrangements based on modern values of development, progress, science, and technology. Risks are part of everyday life, but within the ideology of modernity where security and comfort are predominant values, they have become a normative expectation and standard. The management of risk points in a precautionary direction whereas taking risks or even addressing them can lead to fruitful and constructive tasks and outcomes. Embracing risk is to a large extent required as a basic component to a new thinking that at many spheres and levels is not only an economic but also a social organization challenge. The neoliberal entrepreneurial and experimental forms of governance from the individual to state levels reflect the resilient framework to climate change that has emerged after Hurricane Sandy in New York City. Climate change, extreme climate events, recovery, mitigation, and adaptation unfold within a framework of austerity measures and economic precariousness. Lorey (2015) in her book *State of Insecurity: Government of the Precarious* refers to the implicit precariousness of the finality of life to which everyday existence is exposed to, as also a component of the precariousness under which the management of populations in neoliberalism functions. The governmentality or management of the precarious according to Lorey occurs within a state of insecurity (2015). Accordingly, governmental retrenchment and privatization produce a space governed and administered in a state of insecurity. The current neoliberal and post-Fordist era produces instability and insecurity as a form of governing mediated by the constitution of a subject that seeks security in a complex, unpredictable, and changing economic, political, social, and ecological milieu. The modern ideology of abundance and stability based on progress and development is then anachronistic when placed within the context of neoliberal practices. Accordingly, forms of government under neoliberalism have adapted to administer policies that depart from the precarious footing of governmental retrenchment and privatization.

Risks usually involve a cost-benefit analysis and the recognition that evaluation of risks is contextual—all actions involve an element of risk. The planet to a large extent is managed for the production of resources and governed for the containment of risks (Moore 2016). In the paradox of neoliberal governance that produces instability-based austerity policies, risks are managed within an expectation of safety, stability, and security. Likewise, the management of nature to absorb changes is a form of governance where populations are expected to be resilient (Braun 2014). A large degree of flexibility is necessarily applied to the management of populations and ecology within the uncertainty produced by both neoliberal economic and social arrangements as well as those triggered by climate change.

The new uncertainty or the new unknown posed by climate change rests on the inability to fully predict weather patterns. Briggs (2017) refers to abrupt and nonlinear climate risks. Climatic changes are experienced without a full-proof ability to predict the magnitude, scale, and time of occurrence triggered by climate shifting patterns (The City of New York, Mayor R. Bloomberg 2013). Climate upheavals have profound social, economic, and political impacts and direct consequences to resources in general for an increasingly growing human population. In a risk society where future hazards and insecurities rest on the ideology of modernity itself, climate change introduced yet another level of concern and the need for a multi-scalar perspective that acknowledges change as an intrinsic value with which to reckon at the individual, social, economic, and global levels. Moore (2016) argues that the current transition is at a verge that has the potential to change Earth "rapidly and irreversibly into a state unknown to human experience" (1). Above all, human pressures that affect climate and biodiversity have reached a level of unstable biosphere conditions. In this light, the ways of thinking about and making sense of this uncertain global present regarding climate change are borrowed modern ideas of progress and development currently outdated. In a risk civilization a progressively flattening continuity is implied while the subject under neoliberalism is crafted on the ability to absorb challenges, risks, and change with an attitude of adaptation and resilience. Moore (2016) argues that the ideology of modernity based on ideas of progress and development with the fitting cultural values attached to them have generated a scenario of depletion, unlikely to contribute to its resolution. Even as awareness of a global world at both biotic and social levels is generalized, the modern values of progress and development still seem to pervade our cultural logic (Tsing 2000; Harvey 1992).

THE ANTHROPOCENE

The Anthropocene is characterized by an epoch where the atmosphere, lithosphere, and biosphere are strongly impacted by human forces to the point of massive biodiversity loss, vulnerability to epidemics and disease, and toxic environments, among other destructive ecological outcomes (Moore 2016). In this regard, the notion of the Anthropocene has emerged to mark the contemporary era after the Holocene, as one where human impact has altered the geological footprint of humanity, having a geological effect (Crutzen and Stoermer 2000). In the Anthropocene, humans constitute part of the natural history of the planet, as a species that forcefully impacted the earth in deep time for the last 250 years, "a human-created machine that is now unconsciously bent on its own destruction" (Mirzoeff 2014, 213). The Anthropocene era places emphasis on how various forms of human organization and processes such as states and empires, worlds markets, and urbanization, among others, have reshaped life on planet earth. The concept of the Anthropocene as a geological era is coined in an attempt to highlight and qualify the scale and gravity of the situation and to develop a new lexicon to address the changing conditions of environmental transformation. Johnson and Moorehouse (2014) are slightly critical of the construction of the Anthropocene as an era. For them human intervention as a marker of a strata in the geological record is built upon a mythic anthropos as a geologic world maker and destroyer in an evolutionary narrative that redraws human origins and endings in an exclusively geological context without social considerations. The attempt at bringing a social component to strata requires, according to Johnson and Moorehouse (2014), a more distinctive notion of geologic life that challenges the construction of the Anthropocene as an undifferentiated social stratification. Foremost, the Anthropocene nomenclature shows an apparent dichotomy between society and nature which artificially strips the human from nature, and one might add, nature from human. The ameliorating practices of kindness and commonality among humans, nonhumans, and objects are undermined. Notwithstanding, the notion of human in the Anthropocene is problematic. Human action is a force that impacts the geophysical environment forcefully and negatively. Human action impact is similar to that of a nonhuman, nonliving agent such as the forces that shaped previous geological eras (Purdy 2015; Yusoff 2015; Chakrabarty 2012; Dalby 2007).

Considerations about the Anthropocene anchored in the effect of industrialization, progress, and development make useful and important connections that explicitly uncover the multi-scalar level in which disasters affect people, communities, and governments. Along with globalization, the effects of climate change are symptomatic of the Anthropocene, a multi-scalar, multilayered, global process (Tsing 2000). Human history and natural history in the Anthropocene is a complicated formula where human action on nature equals planetary crisis and where human activities have dialectically and unstably transformed the biosphere and planetary time reaching a critical level.

For Holmberg (2007), memory and social practices affect the perception of natural disasters and in the concept of the Anthropocene, world historical processes are not only producers but also products of changes that have brought about a planetary crisis in the web of life. Humanity's stratigraphic imprint will be part of the human legacy left in planet earth to be studied by future geologists (Moore 2016; Palsson and Swanson 2016; Purdy 2015; Yusoff 2015). One might wonder how human memory and social practices might be interpreted from such impacts. The Anthropocene rests on the assumption that human population growth will continue to increase exponentially. Consumer culture will be part of the structural social models of capital accumulation while wilderness and ecological diversity will be increasingly obliterated in the planet. Technological fixes will likely be required and needed to engineer climate and life in a risky, centralized and industrial scale (Moore 2016).

The anticipatory semantics of the Anthropocene elevates social processes to geological ones and with it the age-old humanist distinction between natural history and human history (Chakrabarty 2012). Adeney states, "The end of the liberal human subject is characterized by a biological and individual self and the subsequent rise of a distributed humanity that operates as a geological agent" (2014, 440). Humanity, critically assumed as a destructive force, built an artificial context mediated by interventions in nature based on technology, planning, and use of resources (Johnson and Morehouse 2014). The Anthropocene could be considered a placeholder to account for the eclipse of globalization as a system that seems to have reached the limits of market deregulation from a liberal state that caters to market forces. Further, the Anthropocene provides an opening to a philosophical event that could potentially trigger a profound shift in our understanding of human relations to the planet and open an opportunity to question the very foundations of the frameworks under which humans, humanity, and its affects stand. The Anthropocene has unsettled the tectonic plates of conceptual conventions to pinpoint a more complex and voluminous, stratified understandings of the earth as a planet and the dynamic geo-social arrangements that constitutes it. In this regard, the Anthropocene also offers the opportunity to present the planet itself as constituted and constitutive of the drama of human politics and a key player rather than simply a static, separated background or stage for human action. Therefore, issues of representation within the Anthropocene are necessarily political problems to retake and rework from new and alternative perspectives for a future that would likely be orchestrated with more intensive forms of capitalist exploitation, environmental degradation, the possibility of mass unemployment, increasing food insecurity, resource wars, and reactionary forms of identity politics (Johnson and Morehouse 2014). McLeman (2017) alerts that the exacerbated rapid population growth, decreasing per capita global food supplies as well as increasing per capita water scarcity will trigger large-scale migrations and displacements mostly in the world's poorest countries. Cities in this scenario would become "ungovernable, erode confidence in government institutions, and undermine peace and stability" (McLeman 2017, 100).

Critics of the Anthropocene as an emerging concept that would open possibilities for a different relationship between humans and nature point at the privileged and self-serving stance embedded in choosing the prefix "anthro" (human) to designate an era as yet another example of a human supremacy complex. The naming of an era that rests on the assumption of human exceptionalism and bounded individualism poses problems when it is that overemphasis on human enterprise that has prompted many of the dire climate change challenges we currently face (Haraway 2016). To some, the label Anthropocene seems to imply another self-serving ontology of the world recoded as "resources," "natural capital," and "ecological service" (Moore 2016). Moore strongly argues against defending a "human enterprise" that has leveled and refashioned the planet at the service of that same human enterprise (2016). The concept itself could render a master identity and narrative based on maintaining the earth's environment in a constant drive to further human development. The concept Anthropocene also contains narcissistic overtones that are supported by the objectification of nature as a fetish that prevents the questioning and development of a critical understanding of the normative tendencies on which it is anchored (Haraway 2016). Moreover, the highly questionable privilege of human exceptionalism on which the prefix for the Anthropocene rests is part of a species supremacist civilization. Chakrabarty (2012) develops a cautionary stance toward the assumed universalizing attempt of the Anthropocene concept. For Chakrabarty (2012), the effects of climate change are already mediated by global inequities. Furthermore, Chakrabarty argues that our current understanding of "human" is limited within the context of globalization, the notion of a global citizen, and the challenges that climate change poses. The Anthropocene as a geophysical force denies a human ontological experience that cannot be experienced by humans themselves, as such (Chakrabarty 2012). That geophysical force does not have an experiential substance for a human identity. This provides a conceptual frontier to establish the possibility of a human/nonhuman alternative understanding of social relations that are not limited to humans only but include nonhuman organisms and objects within a more relational framework as an opportunity to think with nature.

MULTI-SCALAR GOVERNANCE, MULTI-SCALAR ECOLOGY

Accordingly, human presence on earth is a force that produces climate change. Its multi-scalar complexity defies simplistic analysis and single rational solutions because it affects many different levels at once. This global structure characteristic of climate change, according to Chakrabarty, needs to bring together a politics of both humans and nonhumans as active partners in an inclusive non-ontological agency (2012). However, there is no approach or consideration yet for the kind of human/nonhuman relations along the ecological footprint of climate change. Nor is there yet a defined set of values attached to what this metric might be in terms of life expectancy, population growth, economic growth, economic redistribution, and/or global happiness. Many alternatives are likely to be necessary in a risk society intensified by globalization and global warming. The inequalities produced by capitaslism will likely increase risks to include the way we negotiate our attachments to nature and objects. The politics of climate change rely on an open space generated by possible new conceptualizations of the human/nonhuman relationship and "the need then is to think of the human on multiple scales and registers and as having both ontological and non-ontological modes of existence" (Chakrabarty 2011, 15). The multi-scalar register of experience implicated in climate change and implied in the Anthropocene is complicated by the coincidence of different simultaneous crises in terms of resources, finance, food, and weather-related human disasters. The distinctiveness of this historical moment requires forging unexpected and creative alternatives and alliances among differentiated actors based on professions and academic fields to inspire new forms of discussions outside the constraining narratives of state, nation, territory, disciplines, and/or human/nonhuman boundaries (Tsing 2000).

As stated by Sparke (2005), humanity has become an urban species that has changed the conditions of an artificially developed existence interconnected with technologies. However, the manner in which we think about governance and sovereignty is still anchored on the administration of territories that are part of an imperial legacy driven by property, territory, and citizenship in spaces of exclusivity. The artificial ecologies developed in the urban predation of environments have altered the biosphere significantly. The assumed externality of nature to the effects of industrial processes irrevocably accentuates the schism between industrialization and deterioration of the biosphere. Security threats to modernity, and for that matter to constituted states, were considered to come from other states, externally, but with climate change that separation from the environment does not hold (Dalby 2007). Industrial production and the displacement of population from the developed landscapes of urbanization are the causes of environmental and climatic changes (McLeman 2017; Dalby 2007). For Moore (2016), the Anthropocene project rests on a rationalized domination perpetuated into, and as, the future which silences the displaced, killed, and enslaved from those territories that have been assimilated into the logic of industrial production and urban development. More recently I might add from the post-disaster landscapes of forgotten populations. In that same spirit, it is unavoidable to also address the notion of Capitalocene to argue that capitalism's central tenets of accumulation based on unlimited growth have produced landscapes of devastation as a way of organizing nature in the form of a resource to be exploited for profit, constitutive of the capitalist world-ecology. In the face of terror and dispossession of Post-Fordism by economic and environmental catastrophe, some like Shaviro (2015) propose a vision of survival and nourishing based on accelerating generosity, abundance, and self-cultivation. His claim rests on a deep critical analysis of capitalism unbounded acceleration toward accumulation and the Capitalocene as the age of capital. It is within this context that global capitalism has produced an unstable social and ecological system that fosters fast, convenient, and obstacle-free depletion of resources (Haraway 2016). The established disorder of capitalism's law of value actually values nothing as it exploits nature and labor, where nothing is particularly valuable enough to escape the impulse of capitalist exploitation. Capitalism driving force toward the sixth wave of extinction is human driven. Moreover, nature, especially within the capitalist economic framework, is relegated to the level of a resource, only. Like labor, nature is exploited as a cheap yet necessary resource for production purposes within capitalism daring accumulation strategy. According to Moore (2016), nature as resources to be exploited is cheapened as inferior and degraded in an ethic-political sense. Food, energy, raw materials, and human life are cheaply obtained within capitalism.

Capital, power, and nature are complicated in the landscape shaped by class, territory, and technical transformations that emerged in the three centuries after the 1450 industrial economic paradigm or the "Great Acceleration" (Moore 2016). Capitalism in this regard has been a driver toward the extinction of cultures, languages, genocide, and biospheric changes that are anthropogenic as accumulation by extinction or Necrocene. According to Moore, "Catastrophism itself has been a form of knowledge situated within the successive ecological regimes of postwar and neoliberal capitalism" (2016, 8). Capitalocene, therefore, is a rationality produced by the accumulation process. The role of the state in this changing dynamic of power and accumulation is crucial to the administration of resources and the management of populations as part of the neoliberal governmentality (Shapiro 2014; Foucault 2008). According to Johnson and Morehouse (2014), Anthropocene can be understood as capitalogenic also, but to coin the label of Capitalocene for an era obscures the opportunity to question the relationship between human and nonhuman, to which the Anthropocene seems to provide ample latitude for acknowledgment, critical exploration, and possibility for change. As such, Anthropocene is not a constituted identity but rather a space for contestation, critique, and reformulation to tackle the conditions and terms of the future of human existence on the planet in relation to other living and nonliving organisms. Instead of the project of the liberal subject that begins and ends in itself, the Anthropocene allows for the recognition of the human as a geological agent that has left a statigraphic record to the extent that such conceptualization extends from the subject to the ecosystem.

POSSIBILITIES/CONCEPTUAL ALTERNATIVES

The conceptual perspective pushed forward by the Anthropocene portents to upend the defeatist discourse that has emerged from environmentalism and the doom landscape of devastation with which it usually portrays the natural world under current economic and ecological arrangements. Instead, the Anthropocene seeks to present an alternative perspective of humans as rather changing, shaping, and transforming the biosphere (Moore 2016). Even though this approach might be considered a "glossy veneer," it seeks to provide room to consider a more creative process for new ecosystems and conceptualizations of anthropogenic biomes (Moore 2016). Haraway further advances the notion that as humans we should become-with nature rather than the prevalent becoming-over nature relationship embedded within industrialization and capitalism. In that spirit, she challenges limiting Cartesian dualist categories such as nature/humans to alternatives that offer the possibility of considering more complex valuations of food, climate, and nature (2016). Progressive themes regarding our notions of ecosystems and anthropogenic biomes include food and climate justice, and possibilities for de-growth or no-growth regarding population and the economy where population, technology, and abundance in wealth become the pillars of considerations about environmental protections that include human and nature on a conceptually even plane. Additional themes include green or renewable energy, inclusive economic development, a gardened planet, and a global noosphere or a sphere of the mind understood as consciousness where our evolutionary development rests on significant interpersonal relationships with the principal notion of love as its driver (Moore 2016). For many, the resulting capitalism's unequal distribution of wealth and therefore capital accumulation and the subsequent devastation of earth resources and pollution require for capitalism to pay the debt. As alternatives emerge, emphasis is placed on zero carbon emissions regarding energy, engineered crops for the possibility of food shortage, preservation of trees, fish, and other life forms, large-scale desalinization systems to provide water, recycling of metals and plastic, along with other products to which production has altered the capacity of the earth to reaccommodate them in the ecological system, and even the regional manipulation of micro-climates would likely be also necessary to provide a balance between population and the environment. This vision includes a horizon of a population size of "10 billion reasonably rich people" (Giddens 2011, 19). Underpinning visions for the future is the assumption that humanity can have the freedom to choose a way of inhabiting earth that is integrated, relational, and balanced with nature rather than a resignation to techno-scientific rationalization of the old industrial paradigm. This integral community life includes beauty, diversity, abundance, and freedom where working "harmoniously together, enhancing and complementing one another, supporting mutual flourishing, respecting distinct identities and appropriate boundaries, and experiencing union in diversity" are understood as necessary (Giddens 2011, 27). This harmony with wild nature provides different sensorial experiences in its diversity, abundance, complexity, and umwelt rather than conversion, extermination, and commodification. There is an assumed notion that humans can be, according to Giddens, good managers of a standing reserve, which could fittingly lay under the entrepreneurial approach of neoliberalism, as a governing rationality where the nation-state administers and manages the population. Giddens questions humanity's totalitarian regime on earth that lacks a critical stance to articulate alternatives on how governmental/institutional and state roles would be different from a capital mode of accumulation in governing populations (Giddens 2011).

For Haraway (2016) the challenge of a new vision rests on the ability of producing systems with no self-defined spatial or temporal boundaries in that they would incorporate the boundless realm of global ecosystems where information and control is also distributed among its biotic components. The creative uncertainty laid by current conditions would gage and rectify the human ecological footprint. This position seeks to embrace our planetary membership by deindustrializing our relationship with the land, seas, and animals and by conferring biosphere with large-scale geographies of vibrant materialities (Bennet 2010). This realm of creative uncertainty in which we can locate the Anthropocene coincides with the prevalence of resilience thinking from within neoliberal paradigms. There is a need for methods of inquiry and understanding established under open-ended and uncertain epistemological modes where political negotiations are not exclusive of humans but include wildlife and nonhuman forms. The Anthropocene as a conceptual "frack" could contribute to modes of understanding based on cohabitation, coevolution, and coproduction with the planet in all its manifestations as forms of dialogue, alliances, and mutual acknowledgment (Tsing 2000). Attempts at devising research methods to understand and act with the planet could benefit from already liberating methodologies such as those produced in the social sciences like participatory action research which has critically addressed issues of inequality and distributed knowledge and questioned the role of the researcher as an expert as well as issues of representation and power in knowledge production. Yosuff (2015) advocates for geopolitical practices that incorporate inhuman time and materialities, and Chakrabarty (2012) advocates for collective geological agency by thinking of the human on multiple scales simultaneously. Lefebreve (2004) articulated an alternative to linear time referenced by natural rhythms in his rhythmanalysis that despite not including the nonhuman element, it does offer a point of departure to alternative temporal logics. Foucault coined the notion of heterotopias, to speak of other spaces that function under non-hegemonic conditions (1984). Heterotopias, or alternative spaces to the circuits of commodification and consumption that might emerge sporadically and on a temporary basis, could serve to better articulate the sacred, forbidden, and outright otherness where experimentation and possibilities for action could emerge. For Foucault heterotopias are constituted by sacred spaces or the carnival. One might add Occupy Wall Street as an example of an alternative space and culture that emerged for three months in Zuccotti Park, Manhattan, New York City. The space contained donated goods organized as a people's library with free books, a kitchen with free food, a tent with free clothing, meditation practices, among many other initiatives that provided a platform for debate and participation in a culture of solidarity. Battistoni (2013) suggests climate change as a condition for all politics and as a unifying framework and common ground for diverse political projects that could serve as a "lever to wedge open cracks of possibility." The prospect that the very possibility of a future is at stake can free our political imagination and open up avenues for action that might otherwise seem far-fetched (Battistoni 2013). Such avenues must necessarily be experimental. Analysis that focuses on reducing carbon emissions without considerations for an alternative model of population management based on the common good and a common world would require a critical analysis of the manner in which things are produced and consumed, and the kind of energy source to provide viable and complete alternatives. In the hegemonic context of the neoliberal state, the state is then a pivotal organizational nexus in the relation between modern territory, nature, and capital accumulation (Moore 2016).

CLIMATE CHANGE AND STATE ROLE

During industrialization the role of the state relied on the management of resources. According to Giddens (2011), after the Second World War, state planning was heavily involved in the administration of energy. At that time coal mining was nationalized, and during the 1950s and 1960s a shift turning to nuclear power was also guided by the state. The Green Party in Germany was one of the first political advocates to highlight the importance of environmental issues in the 1970s. No-growth society in terms of the economy was an important new idea under which they galvanized support. Their distrust of power and the state exemplified what today could be considered a more generalized sentiment. The Green Party's political framework rested on participatory democracy, distrust for large-scale power, and nonviolence (Giddens 2011). But like with many of the conceptual categories inherited from the paradigms of the industrial era that are inadequate to address current climatic change, the notion of the state is also problematic. According to Dalby (2007), the geography of administration in terms of block of space that the nation-state constructs does not correspond to the fluidity of relations constituted in the connections, links, and consequences of the economic and ecological flows across boundaries produced by globalization, neoliberalism, and climate change. Johnson and Morehouse (2014) argue that "neither the global economy (which is an increasingly large part of the ecological system), nor the global ecological context (which is being remade by that economy) fits easily into the administrative framework of the state system" (443).

The state is an important and unavoidable nexus in climate change. Haraway (2016) contends that the state cannot be avoided in considerations about climate change because for her, in order to be effective, the state needs to be transformed. The state is fundamentally a territorial institution (Parenti 2016) and as such implicated in the ecological processes that have triggered climate change. For Shaviro (2014), the modern state is essentially and literally an environment making process. According to Parenti (2016), modern state power develops through environmental transformation, therefore power, nature, and value (as abstract social labor) are thinkable only in relation to each other. Nature is valued in a peculiar way within industrialization and capitalism, as detached from humans in the alienating practices of production that impinge on culture and social relations. Capitalism places a utilitarian value to everything, including nature. Under neoliberalism, as the political contemporary project, the constitution of the neoliberal subject is the historically contingent product of concrete ways in which the government administers Western democracies for the last two centuries (Foucault 2009). According to Johnson and Morehouse (2014), the detached subject is a product of capitalism's detached humanity and the conditions of its existence. The constitution of the subject produces "beings" that identify themselves with that detachment and in this manner the creation of this subject is shaped by the means of its control. This process of subjectification constructs alienated subjects and the nonhuman as a concrete standing reserve of resources to be exploited. The age of the human is the age of the state administering life. Under this multi-scalar effect, the modern industrial state is not only a producer of environmental changes as it develops landscapes and mediates the administration of resources during climatic events, but also choreographs climate mitigation agreements in the international arena. These are all extremely important roles of the state as the link to connect response, mitigation, and a whole set of measures and policies devised to administer populations and ecosystems.

Zamora (2016) states that the "welfare state, like the liberal state, focused on the economy, it is no longer an economy of material wealth, but an economy of life" (69). Zamora adds that anxiety rising from uncertainty and the lack of security is the foundation upon which allegiance to the state and its categories is based, which involves accepting "everything, taxes, hierarchy, obedience, because the state protects and guaranties against insecurity" (2016, 69) Lorey contributes to this argument by stating that the neoliberal state governs based on the precarity of life in a constant state of insecurity. The paradox of modernity regarding state function is one where state ideology functions as a provider of security and its citizens expect security while governing through fear: fear of terrorism, fear of economic collapse, and lately fear of climate change (Lorey 2015).

In this context the United Nations emerges as a supra-national and external structure that involves state representation of different countries to address climate change mitigation. International Climate Agreements have been set in place to generate a collective and global effort to diminish the impacts of climate change. At an international level, climate agreements are yet another level in which resources and expertise are channeled to administer and manage populations in the negotiation toward climate concerns that states/

governments in their representative role undertake out of the multi-scalar, cross-border, and global impacts characteristic of climate changeability. The new world order advanced by Slaughter (2004) rests on the notion of a disaggregated state that actually functions through a set of global networks "rising the need for and capacity of different domestic government institutions to engage in activities beyond their borders with foreign counterparts" (12). Her approach fits within the required trans-boundary character of governance for climate change impacts. Global governance based on global policy networks is an emerging structure to the neoliberal deregulation of the nation-state responsibilities. Rather than states per se, courts, regulatory agencies, ministries, and legislatures have assumed the task of identifying, developing, and implementing policies to address international problems multilaterally (Slaughter 2004). The implementation of trans-governmental and international obligations rests on distilling and disseminating information, cooperation in the implementation of national and international laws, reconciling national laws and regulations, and responding to common problems. The disaggregated state is therefore a result of the increasing need for government institutions to build on their capacity to participate in actions beyond their borders with foreign representatives out of increasingly global political and ecological concerns and necessities (Slaughter 2004).

GOVERNING DISASTER

A system based on cooperation among territories seems rather necessary to upend the military model generated as part of an historical imperial legacy built upon national political boundaries unfitting to the climatic across boundary security disruptions. Increased violence across national boundaries reduces access to and quality of natural resources to sustain livelihoods that along with climate change undermines the capacity of states to provide opportunities and services to their populations. These direct and indirect impacts of climate change on human security may in turn increase the risk of violent conflict (Barnett and Adger 2007). In this regard, shared environmental problems could trigger social and political changes to promote peace rather than violence. The management of environmental problems could lead to building confidence, trust, and peace between areas in conflict (Dabelko 2006). International conflicts in general and those that directly deal with climate change in particular tend to be managed by using the military. Johnson and Morehouse (2014) state that the military has a very useful role to play in alerting policy makers to the dangers of climate change—and in some circumstances they have equipment that is useful in responding to disasters. But turning environmental change into a threat to national security and impetus to control migration of the poor and marginal after a disaster will undoubtedly produce a violent dystopia (Dalby 2013). The challenge remains on how to handle and manage risk. During Hurricane Katrina, in New Orleans, Miller and Rivera (2008) observed that even though in risk societies the trust in systems that function to protect societies from disaster, doubt, and distrust often emerges based on previous experiences when those trusted systems did not perform as expected. For certain segments of the population the failed protective governmental function based on the high level of insecurities experienced previous to the disaster increased the risk and distrust. In the case of Hurricane Katrina, the disconnect from nature that is prevalent in society and the lack of place-based knowledge produced an over-reliance on socially constructed systems that fails to address the destruction from natural disasters effectively (Miller and Rivera 2008).

ENVIRONMENTAL SECURITY

Environmental security addresses the connection between climate change and the mode in which it is managed, administered, and thought about. The precariousness of extreme climatic events leaves populations at the junction of precarious political and economic conditions. Risk in societies also includes exposure to the insecurity of climatic events that in the United States falls within the organizational scope of Homeland Security, in addition to terrorist man-made threats and attacks. The issue of vulnerability in the discourse of environmental security has been addressed by the United Nations Office for the Coordination of Humanitarian Affairs (OCHA) in reference to slow-onset emergencies or those environmental emergencies that are slow and gradual in showing effects but remain constant and that usually involve a multiple set of causes and effects that often trigger conflict at multiple scales, such as droughts and sea levels rise (2011). Slow-onslaught is different from, for instance, a one-time emergency such as a hurricane or tsunami and therefore requires a different approach to provide the necessary assistance. Therefore the notion of vulnerability requires a different optic under slow-onslaught conditions that pose a different set of challenges to traditional emergency management models to focus more on mitigation, preparedness, and adaptation in addition to recovery.

According to OCHA (2011), vulnerability refers to "the conditions determined by physical, social, economic, and environmental factors or processes which increase the susceptibility of a community to the impact of hazards" (4). In their understanding of vulnerability in the livelihoods for those affected by slow-onset emergencies, developing coping strategies to face gradual change is necessary. From this perspective, humanitarian agencies are challenged to prepare a response to the complexity of multiple hazards and the constantly changing approach required to deal with both, one-time extreme weather impacts and slow-onset, gradual climate change emergencies.

Barnett (2003) states that security is the implicit acknowledgment of vulnerability and as such to assess security there is the need to consider risks, loss, and recovery. Security, vulnerability, and risk are socially constructed and prone to be subjectively interpreted, ideologically charged, and difficult to determine objectively. Security is usually associated to important and vulnerable social bodies such as the nation (national security), basic human needs (human security), income (financial security), and property (home security). The social construction of the process that discursively securitizes vulnerable referent objects and defines specific risks is a political one (Buzan et al. 1995). However, a false distinction is usually made by placing the focus of environmental problems on environmental change as the cause of conflict rather than on economic insecurity and precarity. This false distinction places environmental threats as exogenous devolving into blaming unspecified "others" as responsible for the conflict, a tendency that obscures the social and political domestic causes of environmental problems. The problematic of this misconception is that environmental change is emphasized as the cause of violent conflict rather than the economic conditions of human insecurity. Environmental threats are constructed as exogenous to the state as opposed to domestic political and economic causes of environmental change.

At an applied level, Barnett (2003) advocates for a change in the discourse around environmental conflict and climate change to one focused on security. From his perspective, addressing climate change as a security issue provides the flexibility to integrate and link human security at a local level, with national security and global or international levels of response to environmental change. This framework also allows for the integration of adaptation and mitigation efforts that are basic components of security to climate risks. The emphasis on security, however, should avoid the risk of a militarized response as a sovereign response to a foreign issue and instead one that is increasingly emerging as a problem of the global commons. Many advocate for a reconstitution of a global commons symbolic of the transnational character of climate issues and concerns, as a way to warrant the protection of the environment for all and consider the environment as a human right. Rapid population growth and environmental degradation aggravate conditions of inequality. According to Goodman, Neo-Malthusians consider that population growth and decline in food production and energy resources will trigger wars and civil disobedience as food, clean air, fuel, and farmland face scarcity (O'Callahan 2012). Environmental security is then a recently developed concept to address environmental threats to individuals, communities, and nations as well as the effects of human conflict and international relations on the environment, as these are transnational problems. Scarcity is socially, politically, and economically constructed and as such strongly linked to population growth, environmental degradation, and unequal distribution of resources that adversely affect vulnerable individuals and groups. Peluso and Watts (2001) propose that in order to fully understand the complicated relationship between the environment and violence, it is necessary to also understand the political economy of access and control over resources at both the local and international levels. Issues of production, labor, and distribution of resources are core issues that affect violence over environmental resources. In both the defense strategy and the environmental community, unstable situations could emerge due to lack of resources. Therefore, there has been a convergence between national security, environmental protection, and policy development understood as environmental security (O'Callahan 2012). Data, expertise, and plans have been developed to respond to environmental problems and threats in the national security community. Arguably, national security threats such as terrorism are events similar to climate events in that they both require preparedness strategies not only for one-time terrorist and climate-related emergencies but also for the slow-onset of climate change. According to O'Callahan (2012), based on an interview to Sheri climate change can be considered a multiplier of instability. Their example of Pakistan illustrates this multiplier effect in the case of Pakistan's fragile political situation, a base for terrorism, a strategically important location for the United States, the short-term threat of flooding and the long-term threats of water and natural resources shortages. Environmental impacts multiply threats in addition to the political dynamics of a specific locale and the need to develop appropriate national security planning to indicate the potential for increase stresses for the population affecting poorer neighborhoods and parts of the world more so than developed countries (O'Callahan 2012; Giddens 2011). A global perspective is then necessary to understand the combination of political, social, economic, and environmental issues for better planning and preparation. Homeland Security in the United States developed around protecting against a terrorist threat but evolved to include responses to natural disasters. The preparedness aspect to terrorist threats has some similar components under the administration of environmental security. Homeland Security leads the first response from the government to terrorism and natural disasters.

Looking solely at the impact of hurricanes, the Congress of the US Congressional Budget Office (CBO) estimated that about $23 billion is the annual damage based on sea levels, frequency, population residing, and per capita income in coastal areas (2016). That figure has certainly increased during the 2017 Hurricane season with the impact of Hurricanes Harvey, Irma, and Maria that battered the South of the United States and the Caribbean. The Homeland Security National Response Framework (2013) seeks to address climate events within a framework that is scalable, flexible, and adaptable. This framework includes coordination structures that address the "planning, public information and warning, operational coordination, critical transportation, environmental response, health and safety, facility management services, fatality management services, infrastructure systems, mass care services, mass search and rescue operations, on-scene security and protection, operational communications, public and private services and resources, public health and medical services, and situational assessment" (9). The specific approaches to climate change events will be more directly addressed in chapter 3 about preparedness and adaptation. However, it is important to highlight how Homeland Security becomes, within this initiative, a governing entity in the provision of security by administering and planning the delivery of resources for climate-related events. The administration of resources is organized to include many governmental, private and public, community- and individual-level actors to manage populations in terror, climate, cyber, and even pandemic events. Homeland Security has a crucial role in administering and coordinating efforts as an almost supra-body that connects all aspects concerned with the population in an organizational command structure.

THE GOVERNANCE OF AID: THE US CONTEXT

The Homeland Security responsibility rests on the integration of response core capabilities and mission areas that include prevention, protection, mitigation, and recovery in terms of state-to-state assistance. The different groups and chains of command to administer response to climate events include individuals, families, and households, communities, nongovernmental organizations, and NGOs (Homeland Security 2013). Accordingly the following groups and organizations are included under the organizational framework of Homeland Security: The American Red Cross, National Voluntary Organizations Active in Disaster (VOAD), volunteers and donations, private-sector entities, local governments, chief elected or appointed officials, emergency managers, department and agency heads, state, tribal, territorial, and insular area governments, states, State Homeland Security Advisor, State emergency management agency director, national guard, other state department and agencies, tribes, territories/insular areas, tribal/territorial/insular area leaders, federal governments, secretary of homeland security and other departments such as agriculture, health and human services, Environmental Protection Agency or US Coast Guard, Federal Emergency Management Agency (FEMA), Attorney General, Secretary of Defense, Secretary of State, Director of National intelligence, other Federal Department and Agency Heads (Homeland Security 2013).

In the event that a climate incident of more severity and magnitude exceeds state resources or Federal government, the Robert T. Stafford Disaster Relief and Emergency Assistance Act is requested by the governor (FEMA 2013). The FEMA response stages for climate event impacts include planning, public information and warning, operational coordination, critical transportation, environmental response/health and safety fatality management services, infrastructure systems, mass care services, mass search and rescue operations, on-scene security and protection, operational communications, public and private services and resources, public health and medical services, and situational assessment (CBO 2016). An emergency declaration and if necessary a major disaster declaration are issued to involve a proactive response to catastrophic incidents and a Federal Response and Assistance available without a Stafford Act Declaration. The Federal Departments and Agencies acting under their own authorities are the next in the line of command and Federal-to-Federal Support is established if necessary. Emergency Support Function (ESF) Coordinators from the Department of Homeland Security and the Federal Emergency Management Agency (FEMA), Primary Agencies, Support Agencies, ESF Activation Non-Stafford Act Coordinating Structures, National Response Framework (NRF) FEMA Support Annexes, and NFR Incident Annexes are involved. Among the Federal Operation Centers also participating are the National Operations Center, National Response Coordination Center, National Military Command Center, and the Strategic Information and Operation Center (CBO 2016).

Even though this emergency assistance infrastructure describes in detail the chain of command to deliver aid, a more critical account of the actual delivery in the case of Katrina exemplifies the challenges presented by such framework, as this is an example of a thick bureaucratic web that might not be fitting for the expedient delivery of aid needed after an extreme climate impact. According to Miller and Rivera (2008) the political landscape and bureaucracy under which these responses are administered are based on discretionary spending by both FEMA and the Disaster Relief Fund (DRF) whose budgets are compromised in the payment of current emergencies and ongoing recovery efforts and projects from past disasters. The DRF allocates spending on the Public Assistance, Individual Assistance, and Hazard Mitigation Grant Programs. The Public Assistance Program is the largest covering the costs of repairing or replacing public buildings, infrastructure, and utilities, as well as emergency measures, such as removing debris, establishing temporary shelters, and providing emergency power. The Federal Government covers 75 percent of the Public Assistance Program (Miller and Rivera 2008). The Individual Assistance Program involves 25 percent of the DRF spending and includes temporary housing for displaced people, and repair or replacement of a home and other damaged property, medical treatment, and cleaning, moving, or other needs, grants not covered by insurance, and disaster-related unemployment insurance (Miller and Rivera 2008). The Hazard Mitigation Grant Program, which accounted for 6 percent of spending between 2000 and 2013, supports mitigation measures preventative and includes structural and non-structural activities. Structural activities are designed to prevent or reduce the loss of life or property damage from a future disaster that include strengthening facilities to be resistant to future damage, elevating structures, and implementing building codes that are designed to increase structural resiliency. Non-structural activities include community plans, disaster mitigation and flood plans, and disaster warning systems (Miller and Rivera 2008). The Department of Housing and Urban Development: Community Development Block Grant Disaster Recovery Program, Housing and Urban Development (HUD), and the Community Development Block Grant Disaster Recovery Program work together to fund disaster recovery activities (Miller and Rivera 2008). Block grants funds provide more flexibility for the use by State and local governments than with other forms of assistance that have more restrictive stipulations. The Short-Term Disaster Relief, not covered by FEMA, supports immediate needs following a disaster such as relocation payments for displaced individuals and business, provision of security patrols in damaged areas, and restoration of essential services, such as water, sewer, power, and telecommunications (Miller and Rivera 2008). Mitigation activities designed to anticipate and reduce the amount of damage that a future disaster might cause include strengthening houses and other buildings or implementing flood-control measures and non-structural measures such as funding training exercises, preparedness plans, and buyouts of real estate that is prone to flooding (Miller and Rivera 2008). Long-term recovery is designed to rebuild and economically revitalize affected areas by providing grants to homeowners to repair or replace residences, to businesses that have suffered economic damage, and to local governments to fund infrastructure improvements (Miller and Rivera 2008). Other federal agencies include the Department of Defense with large budgets that does not include the Army Corps to fund the repair of military facilities, equipment, and ships, as well as the costs of deployments for relief operations (Miller and Rivera 2008). The Department of Transportation also with a budget in the billions focuses on the repair of the federal-aid highway system and other critical transportation infrastructure (Miller and Rivera 2008). The Small Business Administration (SBA), The Department of Education Fund, and the Supplemental appropriations to the Department of Transportation also relied on large budgets (Miller and Rivera 2008). The National Flood Insurance Program (NFIP), Mandatory Spending was created in 1968 to provide flood insurance to property owners, the main source of such insurance in the United States. Flood-hazard maps developed by FEMA require that participating communities comply with building-code requirements and floodplain management practices set by FEMA for eligibility to flood insurance programs. Property owners in participating communities could purchase NFIP policies through private insurance companies. Hardship and loss of income caused by hurricanes can determine eligibility for benefits such as nutritional assistance for low-income individuals and families, Medicaid for health-related issues, and unemployment insurance (Miller and Rivera 2008). This complex organizational structure responds to the complexity of climate change events and the social, economic, and political context in which these events take place exemplifying the way in which the government administers populations after a disaster. The shape of this organizational structure was heavily influenced by the 9/11 terror attacks and the development of Homeland Security shortly after.

The administration of aid within the complexity of an extreme climate event poses challenges to the role of the state. Miller and Rivera (2008) based on the experience of Hurricane Katrina in New Orleans relate the story under which FEMA was transformed and the kind of largely bureaucratic decisions that resulted in a less effective response to Katrina. From being an agency with relative autonomy prior to 2002, FEMA's director was given de facto cabinet status for nearly eight years (Haddow and Bullock 2003). Congress created the Department of Homeland Security (DHS) and FEMA was incorporated along with 18 other agencies in competition for parts of the same budget, without cabinet status. FEMA is the only agency exclusively responsible to respond to non-terrorist-related disasters. Under Homeland Security the FEMA budget diminished along with personnel and resources transferred to counterterrorism programs. During the aftermath of Katrina, according to Miller and Rivera (2008), the DHS's organizational focus was misdirected and dysfunctional to address the unexpected consequences of a natural disaster of the magnitude of Hurricane Katrina. Natural disaster became a secondary priority to law enforcement, port security, intelligence, border control, immigration, and transportation security within the restructuring of the DHS. According to Miller and Rivera (2008) the thickening of the bureaucratic organization of Homeland Security to which FEMA fell under, delayed the decision-making process to respond. It is well known that the slow and indifferent reaction of the then president George W. Bush triggered widespread criticism and was viewed as a sign of carelessness and neglect over the mostly and largely black population affected in New Orleans' Katrina who did not have the means to evacuate.

The discretionary spending aimed at helping individuals, businesses, and communities for hurricane damage totaled $209 billion since 2000. More than half of that amount was spent for Hurricane Katrina, and one-quarter was allocated in response to Hurricane Sandy (CBO 2016). The loss of tax revenues also factors into the expenses associated with a climate event and it also impacts any reduction in the growth of GDP in the form of delayed tax payments collected from households and businesses as well as increases in tax deductions associated with personal property losses. Federal spending as a percentage of damage from hurricanes escalated from 17 percent before Katrina to 62 percent after Katrina (CBO 2016).

Changes in the annual frequency of hurricane categories will likely increase due to the effects of carbon emissions, related to increase of temperature globally, and the North Atlantic Basin more specifically—as demonstrated by the active 2017 hurricane season. Continuous coastal development along the East and Gulf coasts in the United States is likely to increase hurricane damage simply by the amount of property that is exposed to damage. The Congressional Budget Office approximated increases in property exposure by estimating increases in population and per capita income by county (2016). Climate change and coastal development will occur simultaneously and each factor will compound the increase in expected damage caused by the other. Rising sea levels also affect the increase in expected damage resulting from expanding populations.

KATRINA

Hurricane Katrina was one of the largest storms to hit the Eastern Seaboard in the United States, specifically New Orleans. The 2005 disaster flooded approximately 80 percent of the city's metro area. Over 182,000 homes and public infrastructure were destroyed (Liu et al. 2011). The levees failure was largely the cause of Katrina's most disastrous flooding condition due to infrastructure malfunction. The temporally closest flood event to Sandy is New Orleans' Katrina. According to Nolan (2011) and Liu et al. (2011) in a report six years after Katrina, 29 percent of New Orleans population dropped, the city's housing stock and property values faced chaotic downturns. The most affected, however, were the sick, elderly, underinsured, or the victims of contracting scams. Thirty-three billion dollars were spent in New Orleans' recovery from Katrina (Liu et al. 2011). Some of that money was used to transform four of the city's largest public housing developments into mixed-income complexes, while still a high proportion of renters live in low-cost, often substandard housing (Plyer et al. 2015). Katrina left on its path the challenges of dealing with permanent loss, financial burdens, different jobs, and different communities. According to Giddens (2011),

Katrina produced the largest level ever of insurance claims. Since 1989, totals of more than $15 billion have been submitted in 10 separate years. Katrina produced approximately an 80 percent flooding of New Orleans in 2005, paralyzing and shutting down the entire city. Communications, water, electrical, sewer and gas services as well as port distribution, and transportation infrastructure dramatically diminished their operations. According to Miller and Rivera (2008), the disaster that Hurricane Katrina produced caused severe damages to the social and physical environment and disrupted the economic, political, psychological, and social life of its citizens. Those who did not have the ability to leave were trapped without resources during the impact of the hurricane and many were later displaced to other states. The disaster landscape left in New Orleans by Katrina highlighted the differential impact and effects between the Lower 9th Ward neighborhood and the touristic French Quarter. The level of social vulnerability that the residents of the Lower 9th Ward experienced was significant and a distinctive marker of the distribution of wealth and economic opportunity (Miller and Rivera 2008).

Miller and Rivera (2008) argue that the way in which people are attached to place symbolically mediates their connection to nature. These connections between people and place, they state, were ignored by federal and local governments in the immediacy of the hurricane's impact in the New Orleans region as it was evident in the slow and low response of relief efforts (Miller and Rivera 2008). Seen as racially motivated by many, the militarized response to the aftermath of Hurricane Katrina was considered insensitive, an example of the unfolding approach to natural disasters led by how terrorist acts are handled. According to Miller and Rivera (2008), the "White House used language that placed natural disasters and terrorism on the same ideological plane" (78). In the case of Katrina, the governor at the time, Governor Kathleen Blanco, declared Louisiana in a state of emergency, implementing the state's disaster plan in action to evacuate residents, which is the initial responsibility of city governments for disaster response relief (Miller and Rivera 2008). The governor then mobilized the National Guard and police. This procedure followed the usual bureaucratic structure designed to deliver resources during a climate event. However, in the case of Katrina it "appears that the state preparedness efforts, in effect long before the storm, were not adequate" (Miller and Rivera 2008, 80). Local emergency management directors are responsible for providing training to prepare for disaster response, where emergency management and local emergency personnel are the first response in disaster delivery (Miller and Rivera 2008). According to Congleton (2006), lobbying efforts granted seven emergency relief bills by the Chamber of Congress and the president, three bills to aid families and students in need, borrowing limits for the NFIP, and a reduced tax bill for the 2006 to 2010 period. All these bills were passed within a month. Local government and communities manage the direct delivery funds but the participation of affected communities did not seem to include decisions about "regulatory, zoning, land use planning, economic development or facility site decisions" (Miller and Rivera 2008, 86). The influence of illegal Mexican immigrants as temporary housing workers also affected the demographic composition with likely political consequences. The "Katrina malaise," according to Miller and Rivera (2008), refers to the inadequate preparation, response, and rebuilding efforts of the communities impacted. Disasters, according to Miller and Rivera, are revealing of the uneven resiliency of communities and the uneven allocation of resources for reconstruction (2008).

The unexpected nature by which lack of immediate coordinated recovery jeopardized communities after Katrina is symptomatic of the casual way by which climate change has been gingerly addressed despite decades' long alerts by the scientific community. The failure to expediently address the impact of a hurricane of Katrina's magnitude point at an unfitting "old order" where governmental agencies' experience with previous national natural disasters did not serve as a learning opportunity to change natural disaster

Figure 1.2 The Rockaway Boardwalk after Hurricane Sandy's Impact in 2012, New York City. Photo by J. Nevárez.

delivery approach (Miller and Rivera 2008). The aftermath of Hurricane Sandy in New York City showed a similar lack of preparedness.

SANDY

Hurricane Sandy occurred on October 29, 2012. The trajectory and the largely corrosive salt-water storm surge caused the sort of damaged now widely accepted as an example of the conditions that climate change has created. The uncertain futures of climate change events like Sandy are likely to continue if the rate of carbon emissions into the atmosphere and global warming do not diminish. Sandy unveiled a quickly overwhelmed old infrastructure and a delayed response to areas in serious need. Subways were flooded which paralyzed transportation for at least a week, and electrical outages were common and lasting. Affected areas in New York included largely coastal areas in the Lower Manhattan, Long Island, Brooklyn, Coney Island, Staten Island, and The Rockaways. The New Jersey Shore was also badly affected from sand erosion and flooding. FEMA declared disaster areas along the coastline and well into the interior of New Jersey and 13 counties in lower New York (United States Department of Commerce, Economics, and Statistics Administration [DOC] 2013).

Climatic events like Sandy and Katrina exemplify broad concerns about climate change. Tompkins and Adger (2004) indicate the ways in which climate change will likely be experienced "slow changes in mean climate conditions, increased inter-annual and seasonal variability, increased frequency of extreme events, and rapid climate changes causing catastrophic shifts in ecosystems" (10). The New York State 2100 Report also indicates that some of the new and mostly unknown risks are associated with pathogens, diseases, and pests affecting both human and nonhuman populations. The expected population increase of 12.4 percent within the next 30 years from 19.6 to 22 million will be an additional source of concern in terms of climate change events preparedness and recovery. The population projections will increase city expansion, suburban poverty growth, and an increase of the senior population by a projected 75 percent (American Society of Civil Engineers 2013). Transportation, energy, land use, insurance, and infrastructure finance will likely be impacted by future climate events and population increase in the New York City area. According to Rosenzweig and Solecki (2015) increase in annual temperatures and heavy downpours are expected as well as sea level rise in the New York City metropolitan region. These are becoming trends that are projected to continue and likely worsen in the coming decades due to the increase of greenhouse gases in the atmosphere by burning fossil fuels, deforestation, and cattle farming. Climate hazards will likely increase risks to the population, the economy, and the infrastructure of New York City as evident in the impact of Hurricane Sandy. Rosenzweig and Solecki state that coastal and low-lying areas, the elderly and very young, and lower-income neighborhoods are highly vulnerable (2015).

The challenge of making the connection between a specific one-time event and the broad spectrum of issues that climate change encompasses has been used ideologically to discredit arguments about the accuracy of climate change claims. Both climate change and hurricanes involve complex ecosystem processes, therefore attributing an event like Hurricane Sandy to climate change poses challenges to a linear causal logic of a unique cause to a unique event. However, higher sea levels certainly increased the extent and magnitude of coastal flooding caused by the storm. Since 1900, sea levels have risen more than a foot in New York City. As sea levels continue to rise, coastal storms will cause flooding over a larger area and at increased heights not previously common. The warm upper ocean temperatures are associated with climate change fuel storms. Many different variables affect the development of hurricanes that make prediction difficult; however, information on the frequency of hurricanes seems to indicate that it is very likely that more hurricanes will occur globally and most certainly in the North Atlantic Basin. As difficult as it is to establish direct causal relations between evident climate change conditions in one specific area to a specific climate event due to the large pattern characteristic of climate change, it is very plausible that the loss of sea ice as the Artic warms contributed to Hurricane Sandy's path and intensity. Since the late 1970s the volume of sea ice by the early fall of 2012 decreased 80 percent largely due to the changing atmospheric steering currents or jet streams that affect the increasing frequency and intensity of extreme weather events. The jet stream affected the unusual westward trajectory of Sandy in New Jersey. Further research is necessary to highlight these very likely relations between sea ice reduction, jet streams, climate events, and climate change (Rosenzweig and Solecki 2015).

The unprecedented meteorological conditions under which Sandy happened included a full moon with high Atlantic tides and a system covering a large area affected by other weather fronts that made a sharp turn to New Jersey where it made land fall triggering a storm surge that devastated coastal communities. Hurricane Sandy arrived on the evening of October 29, 2012 and it overlapped with high tide on the Atlantic Ocean and in the harbor of New York. The high tide at the Battery in Lower Manhattan arrived at 8:54 p.m. and the highest surge there was at 9:24 p.m. The combination of these factors—high tide, hurricane conditions, and a full moon—created a storm tide of over 14 feet above mean lower low water at the Battery, and above Hurricane Dona's previous record of 10 feet in 1960. Hurricane Sandy's landfall with tropical storm force winds covered 1,000 miles, more than three times the size of Hurricane Katrina. The low-pressure system and the wind force that pushes against the water that tends to accompany hurricane systems is strongly linked to storm surge and rise in water level. Erosion, coastal flooding, and boardwalk were heavily affected by the storm surge (Rosenzweig and Solecki 2015).

Moreover, there was the unusual path for a hurricane that made Sandy batter the city's shores with extreme intensity. Before Sandy, most hurricanes that approached the Northeast curved east out to the sea. In the case of Sandy, the winds were spinning north counterclockwise running into a high-pressure system from the north that made the storm tilt to the eastern seaboard. Simultaneously, low-pressure system eastward toward the Atlantic coast intensified the storm and rotated the storm toward the eastern seaboard. Sandy, affected by these two systems, turned west, gained intensity, and landed at 7:30 p.m. on October 29, 2012. Sandy reached New Jersey with maximum winds of 80 miles per hour (Rosenzweig and Solecki 2015).

The storms angle and winds generated a massive surge that directly hit the southern coasts of the New York City facing the Atlantic and the Lower New York Bay, South Queens, Southern Brooklyn, and the East and South Shores of Staten Island. The water inundated beaches, boardwalks, and bulkheads. Seventeen percent of the total landmass of the city, 51 square miles of New York City flooded. According to the FEMA Sandy overpassed the mark of 33 square miles predicted for the "100-year flood" that used to have a 1 percent probability in any given year. Sandy's storm tide caused flooding that exceeded the 100-year floodplain boundaries by 53 percent citywide. In Queens and Brooklyn, the area flooded was more than twice as large as the FEMA floodplain maps. The waves were reported to be 12 feet or more in many areas (Rosenzweig and Solecki 2015).

While Sandy affected neighborhoods all across New York City, the storm hit five coastal areas particularly hard—the Brooklyn-Queens Waterfront, the East and South Shores of Staten Island, South Queens, Southern Brooklyn, and Southern Manhattan. Three of the five areas (the East and South Shores of Staten Island, South Queens, and Southern Brooklyn) were directly exposed to storm surge and destructive waves along the shore, and all experienced widespread inundation. Physical and economic damage was extensive and long lasting. Hurricane Sandy was the worst natural disaster ever to hit New York City. Forty-three people died and many more were affected by flooded roads, the subway system, and electrical facilities. Six thousand and five hundred patients had to be removed from hospitals and nursing homes, 90,000 buildings were located in the flooded zone, 1.1 million children were unable to go to school for a week, approximately 2 million people were affected by power outages, 11 million travelers affected daily, and a total of $19 billion in damages was recorded (Rosenzweig and Solecki 2015). The transportation network in the city stopped working and power outages were experienced in large part of the metropolitan area. The vulnerability to which New Yorkers were exposed after Sandy was an alert into the possibility that storms of such magnitude could devastate the dense environment of the city and the awareness that such storms are expected to increase in frequency and intensity due to climate change (Rosenzweig and Solecki 2015).

The damage caused by Hurricane Sandy affected buildings, utilities, telecommunications, and parks. The hurricane also posed challenges to insurance, access and distribution of liquid fuels, and healthcare provision. Electric and petroleum infrastructures are critically interdependent with other infrastructures such as water, communication, transportation, food supply, and private-sector supply chains (Haraguchi and Kim 2015). The damage caused by Sandy to houses and buildings was severe and widespread. Many buildings were pushed off their foundations by the storm surge and floodwaters or the walls collapsed. Personal possessions and electrical systems were heavily damaged. Approximately 800 buildings in the five boroughs of New York City were tagged by the Department of Buildings for being structurally damaged or destroyed. Others were recorded as damaged containing nearly 70,000 units. Storm-related fires destroyed over 100 homes and businesses, mostly electrical in nature from the combination of water and electricity. Almost 100 hotels were affected by flooding and power outages for at least two to three days as well as airports, especially La Guardia and JFK (Rosenzweig and Solecki 2015).

New York City utilities, especially electric utilities, were heavily affected by the storm since some of the most important utility infrastructures are located on the waterfront. Parts of Brooklyn and Lower Manhattan lost electricity a day prior to the storm due to preemptive measures taken by Con Edison (one of the largest utility companies in the city) in order to protect the equipment and reduce the time without service. When many of the city's substations in the southern part of Manhattan were affected, almost all areas south of the Empire State Building lost power. Approximately 120,000 customers lost power and power lines downed by strong winds affected another 390,000 (Rosenzweig and Solecki 2015). Wireless, cable, and Internet services also experienced outages. Charging cell phones was also a challenge for customers who lost electrical power. Many business and companies set up charging stations in areas affected. Flood damage to critical facilities interrupted landline and Internet service in Southern Manhattan, Red Hook, and the Rockaways. Some providers used the opportunity to substitute for new equipment ending up causing major delays in service restoration. Sixty percent of commercial buildings in lower Manhattan did not have fully restored service 60 days after Sandy when Verizon decided to replace copper cables with fiber optic (The City of New York 2013b).

All parks were closed by the Department of Parks and Recreation the day before Sandy and they stayed closed for three days after the storm's impact until they were inspected and debris and tree branches were removed with the help of many volunteers. Approximately 20,000 street and park trees were downed or damaged by the storm in the city (The City of New York 2013b). Beaches and waterfront park facilities were affected by erosion and coastal flooding in the Rockaways, Coney Island, and the East Shore of Staten Island (The City of New York 2013b).

Most New York City property owners did not have appropriate flood insurance when Sandy hit, mostly because more than half of the affected buildings were outside FEMA's floodplain maps. Many gas stations did not have the electrical power to function properly but also the severe destruction of New Jersey and New York City waterfront infrastructure was an obstacle to the delivery of gas to New York. Refineries, storage terminals, and pipelines were heavily impacted, delaying delivery for many days. Hess Port Reading Refinery in Port Reading, New Jersey and Bayway Refinery Phillips 66 in Linden, New Jersey were shut down (Haraguchi and Kim 2015). Less than 20 percent of gas stations were able to sell fuel one week after Sandy hit (Haraguchi and Kim 2015; The City of New York 2013b). Regarding healthcare, six hospitals had to close—four in Manhattan, one in Brooklyn, and one in Staten Island. Nearly 2,000 patients were evacuated by the City's Office of Emergency Management. Many hospitals that were able to open had a hard time to meet the needs of incoming patients since they were largely operating based on the staff's efforts in pumping water outside the facilities, using lobbies as inpatient rooms, and used gas from cars to keep generators running to provide electrical power (The City of New York 2013b).

The transportation system was also heavily affected causing highways, roads, railroads, and airports to close because of flooding. The six East River subway tunnels that connect Brooklyn to Manhattan were flooded as well as the Steinway Tunnel that links Queens and Manhattan via the 7 train, the G train under Newtown Creek, the Long Island Railroad and Amtrak tunnels under the East River, and the PATH and Amtrak tunnels under the Hudson River (Haraguchi and Kim 2015). The South Ferry in Manhattan suffered major damage along with the viaduct connecting Howard Beach, Broad Channel, and the Rockaways. Private ferries, Staten Island and East River Ferries were also shut down stranding many passengers. The halt in subway service stranded approximately 5.4 million weekday passengers (The City of New York 2013). The transportation systems shutdown impacted about 8.6 million daily public transit riders, 4.2 million drivers, and 1 million airport passengers (Haraguchi and Kim 2015).

In terms of water, those who lived in high-rise buildings' upper floors experienced power outages and were unable to obtain it. A fire in Breezy

Point in Queens also affected water distribution in that neighborhood. The storm also had a strong impact on the city's wastewater treatment facilities. Partially treated or untreated sewage was released into the local waterways. Forty two of the 96 pumping stations that moved storm water, waste water, or combined sewage through the system were temporarily out of service because of damage or loss of power (The City of New York 2013b).

The cost of such a devastating event to insurance companies and the federal government can be measured in billions of dollars: $36.9 billion for New Jersey and $42 billion for New York, specifically (DOC 2013). The disruption of economic activity that Sandy triggered was negative in the anticipated losses in travel and tourism in New Jersey alone (accommodation, food services, retail, amusements, performing arts, and the transportation service sectors) specifically in the areas of Ocean and Monmouth Counties and positive in the likely increase of economic activity from reconstruction projects in New Jersey and New York (DOC 2013). The projected full-time and part-time jobs in reconstruction could reach 281,000 for New Jersey (US Department of Commerce, Statistics and Administration 2013). In New York the state government estimated that the average annual number of jobs supported by a $41.9 billion spending on reconstruction would be 88,000 per year. Economic activity, however, "almost completely resumed within a couple of months after the storm" (US Department of Commerce Statistics and Administration 2013, 5). Other areas affected in addition to Travel and Tourism and Construction were 10,000 manufacturing facilities; Atlantic City Casinos lost approximately $5 million per day in revenue due to Sandy-induced closures, stalled commercial trucks, and the auto industry where in demand for replacement of cars in the flooded areas, and sizable losses in the commercial and recreational fishing (DOC 2013). The insurance companies concentrated payments of $9.6 billion in New York and $9 billion in New Jersey (DOC 2013).

According to the New York City Panel on Climate Change (Rosezweig and Solecki 2015) a large segment of the population reside in the low-lying areas of New York City and in those areas where critical infrastructure to the functioning of the city including natural, economic, and cultural resources are located. These areas are prone to coastal flooding by both warm season tropical storms like Sandy and winter northeastern storms. The evident sea level rise of 12 inches since 1900 likely exposed the vulnerability of low-lying areas in New York City to the approximately 25 square miles flood area that Hurricane Sandy impacted. It is reasonable to accept that sea level rise increases the frequency and intensity of coastal flooding. According to Plan NYC (Rosezweig and Solecki 2015), developed by the Bloomberg administration to tackle climate change impacts to the city, Sandy was an historic event that surpassed the previous 14 hurricanes and northeastern storms since 1900 in terms of the storm surge height, scale and scope of damage. As such, Hurricane Sandy stands as the worst natural disaster in New York City and there was no recovery plan put in place for a climate event of such magnitude. According to Rosezweig and Solecki (2015), just weeks before Hurricane Sandy made landfall, the New York City Panel on Climate Change (NPCC) was passed as Local Law 42. The NPCC is required to meet at least each calendar year to examine recent scientific data on climate change, its impacts and draw recommendations into the end of the century regarding climate change projections. The recommendations are due within one year or at least every three years of the Intergovernmental Panel on Climate Change Assessment Reports publications.[2] Additionally the NPCC has the advisory charge to the Mayor's Office of Sustainability and the Mayor's Office of Recovery and Resiliency (ORR) on community- or borough-level communications strategy development to inform NYC residents about the panel's findings as well as climate change projections (Rosezweig and Solecki 2015).

Figure 1.3 Truck Debris Removal in the Rockaways, 2012. Photo by J. Nevárez.

Hurricane Sandy was a major wake-up call regarding preparedness and recovery to extreme climate impacts. The economic and political context of neoliberalism in globalized cities is critical to understand the tendencies that guide environmental security as deployed by Homeland Security and the management and administration of populations and aid during a natural disaster. The entrepreneurial component that tends to frame practices at different levels from the state to the individual will be addressed in the following chapter as a politico-economic context in the multi-scalar level of climate change governance under neoliberalism.

NOTES

1. https://www.eia.gov/totalenergy/data/annual/index.php#consumption.
2. http://www.ipcc.ch.

Chapter 2

Neoliberalism

Globalization and the Urban Paradox in Climate Change Governance

MANAGING DISASTER UNDER NEOLIBERALISM

Climate change sets forward many confounding issues. Within the context of a post-Fordist economy characterized by free markets, privatization, governmental retrenchment, and austerity measures implemented under the neoliberal framework, climate change events impact at multiple levels. This chapter will look into the neoliberal political ideology that within globalizing cultural forces, a particular paradoxical scenario is produced for the management and administration of populations during and after climate event impacts. Austerity measures and budget cuts have shifted the role of the government from welfare state to retrenchment and disinvestment, where the delivery of services is increasingly managed through privatization under neoliberal policies. Nevertheless, among the many services that the government has swiftly relinquished through privatization, national security still figures as one of its main roles and responsibilities. Human-made and natural disasters still fall within the umbrella of assumed responsibilities that is still considered a priority to maintain a safe and secured population. Despite flaws in the administration of aid during Hurricanes Katrina in New Orleans and Sandy in New York City, the government remains and is expected to be the main actor in the delivery and coordination of aid. Specifically during Hurricane Sandy under the direction of Homeland Security and Federal Emergency Management Agency (FEMA), the governmental agencies were unable to fully absorb the magnitude of the event and its impact to offer effective and timely recovery for an extended period of time, to which citizen organizations and volunteers largely filled the gap. Those spaces that emerged despite a neoliberal framework of privatization, governmental insufficiency under austerity, and income polarization are important considerations for alternatives in the delivery of aid during extreme climate events.

Figure 2.1 New York City Skyline, 2017. Photo by J. Nevárez.

Neoliberalism is generally understood as processes that emphasize financial and labor market deregulation and focus on the preeminence of choice for the delivery of services, privatization of many governmental services and placing responsibility on individuals (Bardy 2016; Braun and Wakefield 2014; Harvey 1992). According to Braun and Wakefield (2014), neoliberalism is appropriate for the Anthropocene as a form of government that administers the flows among the different political, technological, and ecological systems in a process of continuity in a crises-ridden manner. For Brady (2016) neoliberalism is a fitting analytical tool to comprehend the series of interconnections between finance, migration flows, everyday dynamics within state and corporate bureaucracies, as well as individual self-understandings based on the practices and changes occurring at these multiple levels. Neoliberalism is also a rationality, a form of governmentality that carries forth a dominant ideology implemented by hegemonic economic and political groups, as a structural and structuring force and macro-context (Brady 2016; Collier 2012). According to Braun (2014), "government" has become ad hoc assemblages that under neoliberalism and advanced capitalism manage diverse sites and practices in a piecemeal and contingent manner against the backdrop of a dynamic and changing world. The administration of life or biopolitics in this changing context establishes heterogeneous formations of combined diverse elements but also incorporates new knowledge, technologies, and practices in a system where feedback is introduced as part of the operating mechanism of the system itself aiming at its own replication (Braun 2014). Under neoliberalism, governance as the management of life and populations is presented as a posteriori model of continuous combinations of experimental practices to adapt, absorb, and mold in the face of emergent challenges and conditions. The neoliberal management and administration under a constant crises-mode manipulates, intervenes, develops, blocks, and stabilizes. This form of governance utilizes crises justified by knowledge, practices, and institutions in relations of force (Braun 2014). Governmentality then refers to

> the ensemble formed by the institutions, procedures, analyses and reflections, the calculations and tactics that allows the exercise of this very specific albeit complex form of power, which has as its target population, as the principal form of knowledge political economy and as its essential technical means apparatuses of security. (Foucault 1979, 20)

Governmentality in this regard is an appropriate way to study governmental techniques whose application produces specific patterns or issues as a problematic that governments and other relevant sectors of society should address. Governmentality is a concept that also sheds light on how governable subjects fit within specific notions of subjectivity as forms of thought and action that are embedded in technologies of the self. Likewise the rationality involved in governmentality is realized throughout the technologies implemented in the entrepreneurial character of a resilient subject within resilience policies under a neoliberal framework to address crises in a multi-scalar fashion including urban and climatic (Braun 2014). Climate events set challenges to the management and administration of aid in the domain of life over which power has taken control as biopower (Mbembe 2003). Governmentality or the ability of the state to manage its population is put to the test in a disaster like Hurricane Sandy as a function to respond to an urgent crisis. The role of security by the government under neoliberalism could be understood twofold, as providing safety for the population under extreme climatic events and securing institutional practices to provide continuity within a neoliberal approach to government in a society of control and desubjectification.

Mbembe articulates the notion of sovereignty as residing, to a large extent, on the power and capacity to dictate who may live and who must die using as referents the analysis of the holocaust and genocide (2003). This power to govern over life or death can be applied to interpret the governance of extreme climate event impacts. The responsibility to account for each citizen becomes a challenging measure to gage the government's capacity to effectively manage preparation and responsiveness, which for Sandy resulted insufficient and ill equipped. The allocation of resources for aid, the preparation for an extreme climate event, and the recovery process indicate accentuated vulnerabilities in neighborhoods that lack in social, cultural, and economic capital.

In a deeper sense then, the imminence of death could be considered an advisor of sorts, of a life that assumes death and lives with that possibility, as a condition of life. Death holds a great signification as a means of truth seeking. In our climate change present, disasters offer opportunities for reconsideration of future preparedness and mitigation practices. The imminence of death understood as a confrontation with a dramatic change is then a form of agency, a trigger to action. For Virilio living through death in the face of social, economic, psychological, and one might add, climatic catastrophes distinguish our accelerated culture (1998).

State sovereignty is a power dynamic embedded in the management of populations that presupposes the risk of death, of who can live and who can die. This risk has been very much present during the aftermath of Sandy, a risk facing chaos. Besides the usual sensationalization of disaster to which we are constantly exposed in everyday media consumption, climate change impacts pose a distinct and new dystopic scenario against the privileged and protected halo of the urban quality of life enjoyed in a large metropolis, especially reinforced in New York City after 9/11. Urban development initiatives that have catered to the global professional class and tourists alike have stressed emphasis on security and safety especially after the 9/11 attacks, for places like New York City (which is also a confrontation with death and a deeply reflective moment). Mbembe's critical take on what he refers as necropolitics is an attempt at analyzing a politics of death in the polymorphous and diffuse organization that characterizes governmental management by its capacity for metamorphosis within neoliberalism (2003). Applied to the context of genocide, war machines can explain the incidence of mobile segments of armed men that merge or split up with one another depending on the tasks to be carried out and the circumstances. A segmentation of tasks and a plurality of functions are also necessary when dealing with the aftermath of an extreme climate event and one that in the context of the United States falls within the realm of Homeland Security responsibilities. Both man-made/terrorist acts and natural disasters which produce similar scenarios of devastation are under the jurisdiction of Homeland Security. Therefore the direct responsiveness of the state or government to disasters is a continuity of the administrative role of the government in the management of populations. Response and preparedness strategies become imminent. In the militarized strategy deployed as a response to the extreme climate event in the aftermath of Sandy, the National Guard, the Army, and the police assumed a distinct and powerful presence in the territories affected. Even though, possibly necessary and a generalized response, deploying the military embeds a certain lack of flexibility to a crisis. The level of analysis at the upper echelons of the military informed by sophisticated research does not seem to translate into the implementation and deployment of boots on the ground characterized by basic policing practices. During Sandy, governmental response was available but in many instances inefficient and delayed, pointing at the need to rethink and re-strategize response and preparedness to disasters. Predominantly, there is dependence toward the government in terms of responses during crises, including climatic ones. The context in which these responses are crafted is heavily influenced by changes in what is expected from the government, a role that is changing along with the deregulatory and retrenching practices of the neoliberal state. Therefore, the paradox of the state rests on the expected paternal role after a climate event to provide relief while implementing austerity measures and budget cuts to social services that severely hampers recovery.

NEOLIBERALISM

Neoliberalism is a system of strategies and ideological influence across different scales, anchored by globalization's technological advances that provide reach and connectivity to moving people and information. Neoliberalism's economic liberalization and free-market reform have generated "societies susceptible to the flows of capital and to the powerful models, rationalities, and strategies derived from the current project of globalization" (Bogaert 2014, 216). According to Harvey (2005) neoliberalism emerged after the process of capital accumulation produced by industrialization, in an effort to cement profits in what he most recently referred to as accumulation by dispossession. Guided by privatization of services, deregulation and financialization of markets, management and manipulation of crises, and state redistribution, neoliberal policies have cemented capital accumulation by disinvesting in society as a whole. The literature on neoliberalism is abundant and multi-angled. For Brockling (2016), "Neoliberal theories range between moderate and radical, laissez-faire and politically activist positions, between heuristic hypothetical anthropology of a utility-optimizing homo-economicus and evolutionary conceptions of a human nature brought to its economic senses by markets" (59). As a myriad of different perspectives on neoliberalism exist, one could argue that this is representative of the experimental, constantly changing transformation of many structural aspects of the economy and social life that neoliberal policy organically uproots and rearranges. The ability to absorb and adapt to change within the discourse of resiliency that emerged after Hurricane Sandy and now widely incorporated into preparedness, recovery, and mitigation practices is fitting to neoliberal governmentality. One could also argue that neoliberalism is the political expression to capital deregulation and free markets of a globalizing cultural logic based on connectivity and reach. For Springer (2010), neoliberalism is a hegemonic ideology, a policy-based approach to state reform, and a particular rationality of governmentality where the practice of statecraft rests on compromise, calculation, and contradiction. At different scales, from that of the agency of the subject to societal institutions, neoliberalism is reproduced, circulates, and is facilitated, in spite of its variable character. Conceptualizing neoliberalization requires an appreciation of the elaborate exchanges between local and extra-local forces operating within the global political economy (Brenner and Theodore 2002). The analysis of neoliberalization therefore has to recognize different effects where practices highlight contextual specificity and local experimentation (Brenner et al. 2010). Neoliberalism and for that matter globalization, despite being anchored on similar initiatives carried globally, affects places differently and the character of the transformations sought, take on the peculiarities of the experimentation process applied to each locality. Neoliberal policies and programs are characterized by privatization, deregulation, liberalization, depoliticization, and monetarism (Brenner and Theodore 2002, Martinez and Garcia 2000). Capitalist expansion is thus tied to a legitimating neoliberal discourse of progress and development firmly anchored on modern and postmodern ideologies, as yet the most recent expression of a society guided by science, technology, and experts.

The value that neoliberalism places on entrepreneurial initiatives and capacities in a multi-scalar form of governmentality generates the constitution of a subject fitting to governmental policies. Neoliberalism is built upon an image where the state presumptively only oils "the clockwork but not moving the hands or altering the mechanism" by allowing markets to reach their own equilibrium supposedly without governmental interference (Brockling 2016). Making decisions assumes of the individual to be an economic institution whose existence depends on a series of choices. Thus, the individual is expected to be a self-provider and an entrepreneur of their own labor where self-initiative and self-responsibility are crucial components of the knowledge where the neoliberal society is firmly anchored (Brockling 2016). Within neoliberalism the ideological component of entrepreneurialism finds an appropriate fitting into the technologies of the self as a form of self-governance and the governance of others. Appropriated as such, the neoliberal individual cares for the self to establish constant contacts, projects, and activities. Failure is an intrinsic and constant reminder of the claims made by the entrepreneurial culture, global in reach at this point. The entrepreneurial self is a basic component fittingly resting on the governing program of neoliberalism (Brockling 2016). For Brockling, the self is considered to be a social problem and an individual task, not a product but rather a relation of production (2016). Within neoliberalism, the market is the governing principle, not only of economic life and the state itself, but also constitutive of the entrepreneurial subject in relations of competition. For Davies (2014), Uitemark (2005) Swyngedouw (2007), and Jessop (2007), this all-encompassing and multi-scalar neoliberal paradigm is referred to as metagovernance. Metagovernance involves the cultivation of entrepreneurial rationalities combined with new technologies, instruments, and tactics that set collective rules and a particular form of governing. Metagovernance rests on reflexive self-management at a new scale that is less hierarchical and less centralized (Davies 2014). This rescaling and reorganization of the state involves acting strategically to orchestrate rather than command, in spatio-temporal fixes (Uitermark 2005, 24). This form of urban technocratic managerialism rests on a form of governance where policy development, resource allocation, institutional approach, and network participation are based on expert knowledge leadership (Davies 2014). Under neoliberalism, the management of disaster aid assumes a similar approach largely led by technocratic policy initiatives in theory while the application of policies on the ground remains highly bureaucratic.

Brenner and Theodore trace the trajectory of neoliberalism by pinpointing the processes of de-industrialization, reindustrialization, post-Fordism, internationalization, global city formation, urban entrepreneurialism, informalization, gentrification, and an incipient socio-spatial polarization that occurred in the 1980s and 1990s (2010). In the 2000s, urban political economy placed an uneven, contentious, volatile and uncertain character to the urban condition. This took place along with a breakdown of established forms of urban life associated with postwar, Fordist Keynesian capitalism. Neoliberalism refers to "a set of doctrines regarding the appropriate framework for economic regulation, helpful in describing the organizational, political, and ideological reorganization of capitalism that has been imposed through the attempted institutionalization of 'free market' doctrines in specific historical and geographical contexts along with a pervasive naturalization of market relations" (Brenner and Theodore 2002, 102).

However, neoliberal governance is not a finished political program. The different versions are most powerfully held together by the common rationale of "market mobilization" (Brockling 2016, 59). The complexity of the neoliberal project thus involves the role of elite actors and dominant groups that form class-based alliances and have the capacity to articulate and circulate a seemingly coherent program of interpretations and images of the world onto others, a transfer of ownership from the state to private sector or corporate interests for presumably greater efficiency via deregulation, liberalization, depoliticization, and monetarism. Rolled back state capacities of shrinking state interventions are reconfigured into institutional mediations, economic management systems, social agendas focused on urban order, deployment of surveillance, issues with immigration and intense policing. Failures and successes between peoples and their socially constructed realities are reinterpreted and reconstituted to influence forms of knowledge through the "conduct of conduct" (Barry et al. 1996; Brown 2005; Ferguson and Gupta 2002; Larner 2003; Lemke 2002). This is a complex understanding of power through the ensemble of rationalities, strategies, technologies, and techniques that allow the decentering of the government through the active role of autoregulated or autocorrected selves who facilitate "governance at a distance" (Foucault 1991). Neoliberalism advocates for the virtues of free trade, labor that is flexible, and active individualism (Peck and Tickell 2002, 381).

Capitalist globalization to which neoliberalism is its political expression, is a dialectical process simultaneously all-encompassing and highly unstable. According to Davies (2014), neoliberalism is based on entrepreneurial rationalities that actively combines forms of inclusion/exclusion strategies and is coercive. For Harvey (2005), neoliberalism rests on political economy practices where human well-being presumably can best be advanced by "liberating individual entrepreneurial freedoms and skills within an institutional framework characterized by strong private property rights, free markets, free trade based on deregulation of markets, privatization of services, and withdrawal of the state from many areas of social provision resting on contractual relations in the marketplace" (64). Neoliberalism is hegemonic largely as a mode of a self-effacing and self-actualizing discourse with a high degree of flexibility. The temporary contract supplants permanent institutions in the professional, emotional, sexual, cultural, family, and international domains, as well as political affairs. Resting on the creation of technologies of information and capacities to accumulate, store, transfer, analyze, and use massive databases to guide decisions in the global marketplace, neoliberal policy practices impinges into all aspects of everyday life culture. In the worst of scenarios, lack of accountability pervades misplaced decisions regarding budget allocation, to which corruption is a constant concern. This set of circumstances generates a paradox of intense state interventions from government elites and experts in a world where the state, contrary to an image of supposed nonintervention in the market, hence acts as an enabler to capital accumulation.

STATE RETRENCHMENT AND DISINVESTMENT

Disinvestment is part of the restructuring of society under neoliberalism. The retrenchment of national welfare state and intergovernmental processes has imposed powerful new scalar constraints upon cities leading to budgetary austerity in the face of profound socioeconomic dislocation and new competitive challenges. Cities manage a broad array of regulatory responsibilities and socioeconomic risks across interrelated fields of economic development, social welfare, and environmental sustainability. Brockling states, "When life becomes an economic function, disinvestment means death" (2016, 53). Austerity measures are strongly linked to the rollback of institutional practices and they also involve a form of administrative domination evidenced in slum clearances, gentrification, and displacement of working-class people in tactics by municipal authorities who use coercive tools such as demolition, eviction, eminent domain, and rent increases (Watt 2013). Some have referred to the process of rolling back state provision and rolling out less-expensive third sector, and promoting participation in voluntary organizations as a "flanking mechanism" of neoliberalism (Jessop 2007). These mutations have also entailed a number of significant institutional realignments at the urban scale with the empowerment of business-led networks and agencies, interim cooperation and industrial development clustering, development of community-based programs and shadow state initiatives to combat social exclusion, the reorganization coordination and interorganizational local state, development of regional institutions for place-marketing and coordination of intergovernmental agencies (Clarke 2009). These multifaceted processes of local institutional transformation are based on contested and trial and error processes with different experimental strategies that are place specific.

Contestation has emerged out of the implementation of draconian neoliberal policies of austerity measures where social and public services are privatized and the government retrenched in the provision of social services to the "rollbacks" and late-neoliberal "rollout" practices (Clarke 2009). Rollback refers to the streamlining of administrative processes while simultaneously introducing new "rollout" economic management systems as policies based on a social agenda around urban order, surveillance, and policing as forms of governmentality in the management of large populations during constant economic, social, and/or cultural crises (Peck 2001; Peck and Tickel 2002; Springer 2010). These rollback strategies include privatization, deregulation, liberalization, depoliticization, and monetarism (Brenner and Theodore 2002; Martinez and Garcia 2000; Springer 2010). Rampant inequality and gentrification intensify the contradictions of a neoliberal state engulfed in austerity measures with social provisions' reduced budget. Neoliberal agendas that are market disciplinary are further entrenched within the underlying institutional structures of urban governance. The neoliberal ideology of constant change, improvement, and competition involves failure and mistakes as constitutive of the experimental learning process of new policy development.

Within rollout neoliberalism, the powerful contradictions inherent within the current urbanized formation could provide openings for more progressive and radical democratic reappropriations of city space. The crisis in the

Keynesian welfare state will likely deepen the crisis in the neoliberal project and could open up opportunities for both reformist and counter-hegemonic movements. Alternatively, destabilizing neoliberal rationalities, strategies, and technologies that aim at disassembling the neoliberal project could destroy its techniques. One of the key components to the transcendence of neoliberalism is, therefore, the construction of new forms of urban solidarity between and within cities (Clarke 2009). For Harvey (2003), the challenge rests on how local initiatives that exercise a critical bottom-up articulation of alternatives would need to establish alliances outside their locales. For Brockling (2016) the alternative rests on ephemeral moments where progressive actions and political projects meet.

Cities' spatial justice along the distribution of environmental impacts shows a spotty and uneven geography that corresponds with neoliberal austerity measures. For instance, the Buyout Program from which the FEMA offers to buy the properties affected by Sandy as part of the hazard mitigation assistance is unevenly applied. Likewise, many measures set in place to address climate change also unfold unevenly. Some houses are elevated after a new code for flood probe areas was introduced, while others, mainly of those with limited resources, remain as originally designed. The economic recession in 2008 that still lingered in some sectors in 2012 at the time when Sandy occurred, was already an event that required resiliency from the population, especially from young educated adults without job prospects. Resiliency then is the ability to overcome obstacles, to adapt and figure out solutions largely based on trial and error that even though temporary are part of a learning experience. This is a common-sense assumption that becomes complicated when the ground of possibilities is structurally shifting within the context of market deregulation, outsourcing, and a large labor reserve. Maybe not surprising, this is a very fitting ideology to the entrepreneurial spirit of neoliberal tendencies and part of the controversial manner in which the analysis of neoliberalism and policy practices remain contradictory forms of governance situated in the city.

URBAN ENTREPRENEURIALISM: SHIFTING GOVERNANCE

The impact of neoliberal policies in the urban environment highlights the city as a pivotal scale within the neoliberal agenda. The city became an increasingly important terrain of struggle for neoliberal hegemony (Brenner and Theodore 2002). Spatial organization is at once structuring, a site, and a medium for the mobilization of neoliberal political strategies. Cities have become institutional laboratories for neoliberal policy experiments (Davies

2014; Brenner and Theodore 2002). This experimental approach is a cornerstone of neoliberal entrepreneurial ethos and one that in no accidental way also leads climate change policy (Broto and Bulkeley 2012). The rise of "resilience" as a set of discourses, policies, and practices oriented to further insert cities with the neoliberal order, better enables investors to assess risk in urban areas (Coaffe 2009). Moreover, urban neoliberalism is a form of not only political, institutional, and geographic change but also has at its core a means of transforming the dominant political imaginaries as a myth from which people understand the limits and possibilities of the urban experience. In this redefinition of the political imaginary, assumptions about the role of state institutions are rearticulated along with the more general inherited understandings of citizenship, community, and everyday life through a lens of entrepreneurial impulse across scales. The ideological aspect of neoliberalism understood as a singular, ahistorical and uniquely efficient market-led system, naturalizes neoliberalism constituting it as a rather standing-alone

Figure 2.2 Upside Down Metropolitan Life Building Sculpture, Flatiron District, New York City, 2011. Photo by J. Nevárez.

and self-reinforcing discourse. For Clarke (2009) neoliberalism tends to exist in a kind of parasitical relation to other state and social formations such as neoconservatism, authoritarianism, and social democracy, among others. Thus the notion of an almost autonomous, self-regulating market is a productive myth for the restructuring strategies shaped precisely in and through these recombinant contexts. Neoliberalism's evocation of a standing-alone market order is a strong discourse and a self-reinforcing myth. Neoliberalization works through a trial and error experimentation, mostly under conditions of crises and regulatory failures with dysfunctional and disruptive consequences. Therefore, neoliberalism both "exploits and produces socio-spatial difference" (Clarke 2009, 53). Uneven development represents a coevolving and codependent facet of the neoliberalization process (Smith 1996). The rollback face of neoliberalism is integral to the dynamics, logics, and trajectories of the regulatory transformations unleashed by a free-market ideology. Neoliberal rollback strategies with the subsequent struggles over the regulation of capital accumulation and its associated contradictions result in the recombination of inherited and emergent institutional arrangements. Neoliberalization is an open-ended process in post-Fordist Keynesian systems of production and reproduction that places cities as key arenas for neoliberal rollback strategies where innovation, growth, and institutional experimentation are localized. Cities have become critical nodes and points of tension in the evolving scalar politics of neoliberalization in an accentuated manner (Clarke 2009).

At a more applied level, Swyngedouw et al. (2002) analyzed the emergence of new forms of governance in the relationship of large-scale urban development projects in 12 European cities that included Berlin, Copenhagen, Dublin, Athens, Brussels, Naples, Bilbao, Vienna, and Lisbon, among others. Their study focused on the political, social, and economic power relations in the city under the neoliberal urbanization paradigm. Swyngedouw et al. (2002) found that new forms of governing urban interventions in large-scale projects were characterized by less democratic and more elite-driven priorities. An ambiguous impact of these projects on the city at large was found, where local democratic participation mechanisms were not respected or were applied in a very "formalist" way, resulting in new elite power arrangements. However, grassroots movements occasionally managed to impact in favor of local participation with minimal social returns for groups that were deprived. Despite the rhetoric of market-led and private investments, the state acted as leader. According to Swyngedouw et al. (2002) mechanisms of representation formally exerted rarely existed or were co-opted by invitation from key institutional power brokers leading to an inability to identify who represented what and how. This invariably influenced the decision-making processes, the environment and interventions of those included but not of those absent, therefore marginalized (Swyngedouw et al. 2002). The targeting of spaces for development in this refashioning of selective spots for large urban development projects as an urban development global trend allows for the representation of particular social groups as problematic, excluded, marginalized, and nonintegrated. Neoliberal forms of urban governance contribute to accentuate processes of social exclusion and polarization (Swyngedouw et al. 2002). Urban space, within these initiatives, is constituted as an abstract objective for the development of powerfully ideological and exclusionary practices. The way the process unfolds generates the ideal conditions for the consolidation of a neoliberal and market-driven ideology and politics. Social inclusion distribution–based policies and Keynesian demand–led interventions are withdrawn from the government. Spatially targeted social policies and indirect promotion of entrepreneurship, especially "via selective deregulation, stripping away red tape, and investment partnerships" substitute state interventions (Swyngedouw et al. 2002, 551). Urban revitalization is then presented as an opportunity to change economic hierarchies and functions. In this way the search for growth turns urban renewal into a mediated objective and a necessary precondition for economic revitalization. Even though this general trend takes distinct expressions in different cities, project based urban interventions generally involve critical changes in priorities and the rise of a more assertive and dynamic entrepreneurial style of urban governance (Swyngedouw et al. 2002). Closing the rent gap and maximizing profit on the produced revalorization of the developed land seems to guide most of these large-scale urban development projects in Europe. The role of lobbies, family ties, business connections, and forms of "clientelism" become dominant and generate a dual society along the coalition of public/private interests and a growing disenfranchised group. This neoliberal urban development approach generates islands of wealth in an impoverished environment. The city becomes a spotty mosaic of socioeconomically highly diversified and more mutually exclusive areas, with considerable state support or subsidies. This kind of neoliberal "spotty" urban development produces a geography of the city as a patchwork of democratic deficit changing the built environment for the accumulation strategies of a city's key elites, along with a fragmentation and privatization of urban governance. Under this neoliberal framework, the "entrepreneurial city" model moves from government to "governance," from a set of formal protocols and institutional roles geared toward social interests, social dispute resolution, and public choices implementation, to a generally flexible pattern of public decision-making supported by loose associations of individuals in influentially diverse organizations and territories. Cities then have become geographical targets and institutional laboratories for a variety of neoliberal policy experiments, from place-marketing, enterprise zones, local tax abatements, and public-private partnerships. The overarching goal of these policy experiments is to transform city space for both market-oriented economic growth and elite consumption practices, while securing order and control among marginalized populations (Clarke 2009). These measures have also produced, amidst these market-disciplinary urban policy initiatives, other forms of governance. They are significantly realigned at the urban scale to include local economic development policy of business-led networks and agencies based on cooperation and sector clustering, community-based programs aimed at diminishing social exclusion, among other initiatives (Clark 2009).

The entrepreneurial city governance under neoliberalism promotes and normalizes a "growth first" approach to urban development and reconstitutes social-welfare arrangements. Urban policy within the context of the entrepreneurial city anticipates, complements, and in some cases mimics the operation of competitive markets. Neoliberalism not only privileges lean government, privatization, and deregulation but a combination of competitive regimes of resources allocation, skewed municipal lending policies, and political pressure that undermines and tends to limit alternative paths of urban development (Clarke 2009). This produces a neoliberal stagnation in the public sector based on austerity measures and a growth-chasing economic development that is outward looking and replicated. An aggressive posture on the part of local elites and states emerges to actively search for opportunities by monitoring competitors and emulating best practices. Despite its language of innovation, learning, and openness, neoliberalism is associated with an extremely narrow urban policy repertoire based on capital subsidies, place promotion, supply-side intervention, central-city makeovers, and local promotion in the branding of places (Clarke 2009). Economic potential indicators and the administrative capacity of the city are led by funds allocations rather than social need. Therefore, areas with high incidence of poverty and social exclusion are targeted for development justified by concerns about crime, welfare dependency, and social breakdown. Cities are confronted with after-welfarist forms of governance and in many instances confronted with organized resistance to neoliberal policies. Entrepreneurial regimes of urban governance are, therefore, not simply local manifestations of neoliberalism but rather extensions of global trends in neoliberal policies (Springer 2010).

The urban paradox of the city and climate change according to Davis (2010) lies in the understanding that the city is simultaneously the problem (as spatial, economic, and social organization that emerged from industrialization) and its own solution (where more progressive policies toward climate change mitigation tend to be developed). Braun and Wakefield (2014) also point at the fact that the city is considered the most pressing and promising locale for climate change future scenarios as the city is where the future of humanity's social organization and physical design coincide to face the challenges of climate change. As demonstrated by Hurricanes Katrina and Sandy, these were events that placed at center stage relief approaches to differential vulnerability of cities, neighborhoods, and specific populations to unexpected adversity. Cities as integrated socioecological networks of global systems are transformative agents and vulnerable subjects simultaneously. Therefore cities are a challenge to governance when crises are all pervasive and the next disaster inevitable to be absorbed, attenuated, and survived under conditions of austerity and government retrenchment.

Globalization generates a culture of shared practices and beliefs easily found in the urban development patterns adopted, policy measures that facilitate capital accumulation and disinvestment, austerity policies and emergent self-organizing responses to which Occupy Sandy is an example. The thick landscape of globalization within neoliberal deregulatory practices has placed the city at a central location of multi-scalar interventions that are key components to climate change impacts and recovery policies. Especially with regard to governance, cities are critical nodes in the management of climate change risks and mitigation (Vivekananda and Bhatiya 2017). Cities are also critical nodes to address climate change (Braun 2014, Broto and Bulkeley 2013; Corfee-Morlot et al. 2009). According to Broto and Bulkeley (2013) a heterogeneous group of "actors, settings, governance arrangements and technologies involved in the governance of climate change in cities in different parts of the world" (92). The common denominator in these heterogeneous arrangements seem to be what Broto and Bulkeley conceptualized as experimentation in urban infrastructure systems and to a certain extent experimentation in social systems (2013). Experimentation is then a key aspect that could open new political spaces for governing climate change in the city according to Broto and Bulkeley (2013). In this regard, social and technical forms of experimentation in response to climate change are a fitting component of the entrepreneurial character of neoliberal urban development. These interventions seek to understand new ideas and methods to develop the ability to absorb change in the context of uncertainties and vulnerabilities. Broto and Bulkeley found that experimentation on urban climate change responses is a global trend (2013). They also point out that urban centers are considered a vital part of the global response to climate change and that cooperation between countries, different levels of government, and the private sector is required. The notion of climate change experiment, according to Broto and Bulkeley (2013), entails governance niches and new grassroots interventions in socio-technical regimes, and urban laboratories as systems that are part of governance and challenge to socio-technical arrangements. Urban climate experiments therefore are purposive and strategic interventions that provide new forms of learning experiences to mitigate emissions of carbon gases and/or adapt to climate change impact vulnerabilities. They are considered innovative and urban when they are administered based on a preconceived notion of an "imagined urban community." The local, community component to disaster recovery is one of the multi-scalar levels involved in climate change recovery. Therefore they open up possibilities for innovative political spaces for governing and administering climate change in the city. Broto and Bulkeley concentrated on six indicators that contribute to greenhouse gases and concentration of vulnerabilities to climate change in a representative sample of heterogeneous group of cities around the world (2013).[1]

In municipal responses to climate change mitigation efforts, Broto and Bulkeley's research focused on six indicators, namely urban infrastructure, built environment, transport, carbon sequestration, and urban form as well as adaptation experiments (2013). The initiatives studied were led by initiating actors, partners, donors, and supporters from both the public and private sectors. The governance of adaptation and mitigation experiments occurs at a multilevel and multiactor process, therefore involving local agents and transnational municipal networks (Hsu et al. 2015; Broto and Bulkeley 2013; Bierman 2012; Keohane and Victor 2011). Moreover, participation in transnational municipal networks in the form of climate agreements showed a strong correlation with a higher number of experiments in specific cities such as London (Broto and Bulkeley 2013). Local governments tend to lead the urban climate experiments along with the private and civil society actors (Broto and Bulkeley 2013). Likewise, the false split between the local and the global has become increasingly problematic when shifts in governance and policy implementation reconstitute places as players in the economic feasibility of cities.

TRANS-LOCAL PREPAREDNESS AND CLIMATE CHANGE URBAN GOVERNANCE: NEOLIBERAL/ACTIVIST LOCALISMS

There has been a misleading distinction between the global and the local in globalization and on the often heavy emphasis on the "local" that makes it seem as only an appendix of the global. More recent conceptualizations have exhausted the global on contemporary social, economic, and political analysis while the local usually remained as an idealized and unexamined place anchored on an unproblematic and assumed cohesiveness. According to Brenner and Theodore (2002), within neoliberalism, a reorientation of policy priorities has also ranked political identities while the meaning of urban and national "community" is redefined (2002). The new politics of scale, political-economic dimensions of the new localism are emphasized generating spaces of neoliberalism and as such referred as "glocalization" (Swyngedouw 1997; Jimenez-Dominguez 2009). At the juncture where the global and the local intersect, change and resistance are structurally intertwined. According to Braun (2014), the local, as the delineation of the nation-state has changed under market deregulation and neoliberal policies is constitutive of a diverse composition of populations in movement with information overload. The global ecology is an additional layer of interconnectedness between the local and the global in a way different than the cultural impact of globalization. The fast policy of neoliberalism has produced a new localism, a process whereby new public and urban policies are continually generated in localities and circulated in a somehow accelerated manner between localities where policy reform is constant and normalized, with accelerated policy turnover and shortened policy cycles. This interurban networking in urban policy mobility grants expertise, policy models, and "best practices" to be shared among national and international organizations, interpersonal contacts, and fact-finding visits (McCann 2008). Local policy professionals, politicians, and activists resting on globalization's ease of fiscal policy transfer, involve a range of actors and a range of forms as part of urban policy mobility based on interurban networking and interurban cooperation, and virtual connectivity. Policy makers learn from international experts how to convert policies into models that travel by mobilizing international case studies into local politics to win over opponents, polishing discourses, and framing debates by various media (Clarke 2009; McCann 2008; Ward 2006). The role of academics in this arrangement, along with other professionals, can be seen as mediators and translators within the expert class of policy shapers. Thrift (2005) has a less enthusiastic perspective and defines soft capitalism as a set of contemporary economic practices that assume a permanent state of turbulence and vulnerability that is best survived through small organizations composed of fast subjects and reactive managers. Contemporary policy repertoires tend to be narrow and exclusive, leading to policy importation and churning because variations on the same theme continue to fail and generate demand for replacement policies (Clarke 2009). In a slightly different manner, Corbridge (1998) offered a counter-critique of post-development by pointing out that poverty is not mainly defined based on census estimates but that in fact basic needs do exist, that local culture could also have a dark side as it does for women in many instances, that technology has a benign side where it could be labor and life-saving, and that development in some instances has been approached critically and reflexively as a dilemma. Cooperation has been found to take multiple forms in these policy networks including information sharing, cost pooling, joint lobbying, collaborative planning, joint projects, and the dissemination of "best practices" (Brenner 2004; Graham 1995, Leitner and Sheppard 2002; Leitner et al. 2002; Clarke 2009). Braun (2014) points at the new and novel forms of technological modes of management to overcome climate crisis by modulating efficient practices to our relationships to carbon in global urban systems. The larger natural environment is the milieu where cities are embedded and contextualized. The natural environment therefore is a historical product as much as the city itself. Our relationship to nature and ecosystems is also a product of the way we comprehend and interact with it, under the complex historical, economic, and cultural circumstances to which it relates. Resilient strategies or "resilient urbanism" to the scale of the city requires the modulation and management of natural processes to include the nonhuman world (Braun 2014). Resiliency emerged as a predominant component of the preparedness and adaptation discourse to be addressed in the next chapter. Currently it serves as an overall approach that distinguishes neoliberal practices with an experimental outlook to policy development and implementation.

The trans-local character of climate change policy practices within municipalities and/or cities has been the major focus of study for Kemmerzel (2016) and Zimmerman (2016). Kemmerzel studied the multilevel governance structures and the trans-local climate change policies in 64 German cities with more than 100,000 inhabitants each. Cities are privileged sites for access to best practices regarding the trans-local dimension of local climate policy to address solutions or collective interest support on such policies. Trans-local activities provide local actors the opportunity to showcase their preferred policies. Trans-local in this context refers to membership in municipal networks that are transnational in character and to the implementation of network goals toward agreed climate change policies. The strategic response to urbanism considered in the adoption of climate change policies involves a polycentric and horizontally oriented collective action within multilevel governance arrangements. Examples of climate change mitigation policies in this context are concerned with the rate of carbon produced or solar energy use as a climate change stressor. The climate change mitigation strategies are embedded in current urban sustainability issues where the environmental adoption score of the city is checked (Homsy and Warner 2015). The aspects considered in the multilevel governance are intra-administrative capacities such as availability of professional staff, population, and population change (Homsy and Warner 2015). The metro status of a municipality, fiscal constraints, and civic capacity based on education and prosperity were considered.

The importance of trans-local action, Kemmerzel argues (2016), is that it provides the opportunity for local actors to support and justify local measures for climate change mitigation. I would argue that trans-local actions are necessary due to the global, multi-scalar level of governance that climate change impacts required based on the trans-local or across national boundaries that link climate in general. These have direct consequences for environmental policy making, particularly in cities, where most of the world population resides. Climate change mitigation involves local policy directed by decisions, incentives, and approaches from higher levels of government as well as horizontal cooperation among different actors and communities (Kemmerzel 2016). Actor orientations regarding municipalities' governance relations in the multilevel governance of climate change are not bound to a particular scale in its vertical dimension (Betsill and Bulkeley 2006). The horizontal dimension, argues Kemmerzell (2016), includes city networks and the transnational municipalities networks that in the European context translates to Climate Alliance, Energy Cities and Cities for Climate Protection, C40 Cities Climate Leadership Group, and the working group under Climate, Air Quality, Climate Change and Energy Efficiency (Kemmerzel 2016). In nearly all the German cities included in the study, the members of the administrative staff act as policy entrepreneurs and try to implement innovative policies (Kemmerzel 2016). Kemmerzel found that the involvement of cities in vertical and horizontal trans-local activities is a sufficient condition for the adoption of innovative climate protection policies and that trans-local action has a distinctive effect on the adoption of innovative climate protection policies. For Krause (2011) education, household income, and political representation are strongly related to civic capacity pending on the wealth or education of a metropolitan region. Zimmerman, Boghrat, and Weber (2015) studied the climate change policy knowledge as a new competency and the ability to communicate policy information in three German cities Munich, Stuttgart, and Frankfurt. According to Zimmerman et al. (2016), "Economic, political and ecological interests in socio-technical transitions towards low-carbon cities are mediated through an intricate architecture of knowledge forms, epistemologies and organizational relationships" (2). The emergence of new tasks attached to the implementation of climate change policies generates new administrative units in the form of "municipal climate management" with specific expectations on how to act. These new tasks and expectations along with the corresponding local policy instruments are increasing in number. The municipal administrations' organizational knowledge structure and management is bound to also change with the new climate change policy instruments as these fit or are resisted in the administrative behavioral routines. The governance of climate change problems, according to Hoppe (2011), falls within at least two arrangements. The first, closed and institutionalized networks are largely technocratic and prefer expert knowledge that is bureaucratic and usually "invited" participation. The second is a more open and participatory citizen model with consideration to multiple scientific perspectives where the knowledge order is more pluralistic and problem solving in a process that develops incrementally. The desirable alternative would include elements of the latter in a multi-scalar governance form where the city and municipal levels of organization offer service, advice, and conceptual aid. Zimmermann et al. (2015) argue that an epistemological shift occurs in the implementation of the diversity of factors involved in climate change policies when the numbers of actors, new ways of knowing, new technologies to disseminate information, and assessment tools emerge along with the increase of the different governance arrangements available. For instance, issues produced with the heat island effect, heavy rain, floods, heat, and drought, the diminishing number of domestic species and the appearance of invasive ones require climate change policies that are different from those administrative and organizational arrangements in municipalities facing these issues. In the context of New York City, at the community level, the necessary combination of bottom-up initiatives was coordinated by a group of community organizations that stepped into securing the needs triggered by Superstorm Sandy's impact, including the Occupy Sandy's effort which influenced the development of preparedness and recovery policies at the city level. The mayor of New York City and the governor at the time have also pushed forward plans and initiatives largely based on professional expertise or knowledge from policy entrepreneurs and trans-local knowledge transfer. In Frankfurt, Germany, most of the climate change policies are developed by knowledge policy entrepreneurs in government offices with limited knowledge transfer who did not sufficiently broaden the rising climate change mitigation strategies concerns (Zimmerman et al. 2015). Viewing climate change as an issue for a stand-alone department such as Energy Agency or Department of the Environment in the case of Frankfurt, Germany, limited the level of collaboration with other climate change relevant departments such as planning or transportation (Zimermann et al. 2015). In Stuttgart, dealing with an air corridor that causes heat island effect and poor air quality, their climate change policy strategies implementation happened within empirical experiments where varied innovations and strategies are put in place (Zimmermann et al. 2015). In Munich, climate change does not seem to be a clear priority in their urban politics (Zimmermann et al. 2015). The epistemological order regarding climate change policy in Munich risks fragmentation by being in constant flux by its decentralized and open implementation of climate change mitigation projects. There is a plurality of knowledge production sources in Munich with weak decentralized links among different departments, according to Zimmerman et al. (2015). The risk of Munich approach to climate change strategy implementation is that it might be led as policy-based evidence rather than evidence-based policy (Zimmerman et al. 2015).

With regard to the kind of knowledge necessary to deal with climate change conditions, Zimmermann et al. (2015) found that in contrast to Stuttgart, Munich, other cities in the study depend on external expertise to complement their own administrative expertise. Knowledge transfer in their view is utmost important to step up climate change policies. Zimmermann et al. concluded that climate change policies are institutionalized differently in their three case-study scenarios (2015). Frankfurt was at the forefront of climate change policy implementation in the 1990s, Stuttgart does not seem to have accomplished a specific form of organizational change, and Munich has a decentralized and networked coordination as part of its implementation model. Zimmerman et al. (2015) suggest that it is highly desirable to develop a stable epistemic community among different concerned actors in an open pluralistic arrangement that encourages participation from the bottom-up but also allows for a robust technology transfer and as part of a larger strategic urban development process with climate change policies at its core. Referred to as policy experimentation by Broto and Bulkeley (2013), 79 percent of the 495 urban climate change experiments studied in their international database occurred after the Kyoto International Agreement in 2005. International agreements therefore tend to increase awareness and policy actions toward climate change mitigation and adaptation.

EMERGING INTERNATIONAL CLIMATE GOVERNANCE: CLIMATE AGREEMENTS, COALITIONS, AND COLLABORATIONS

Governments around the globe have participated in climate mitigation agreements and it is generally understood that these are necessary steps to diminish negative climate impacts. According to the US Congressional Budget Office (2016) atmospheric temperatures, sea-surface temperatures, water vapor, and wind shear are some of the factors that affect how often hurricanes happen, and these conditions tend to be quickly responsive to changes in carbon emissions. Even though a strictly direct correspondence between carbon emissions and frequency of hurricanes is difficult to establish in a linear logic, reducing carbon emissions in the atmosphere is a strategy generally supported by most countries' governmental state and/or city representatives, including those who are among the largest polluters like the United States, China, India, Brazil, and South Korea. More recently, the current administration in the United States relinquished their participation in the Paris Climate Agreement, but cities have assumed leadership in continued participation in de-escalating carbon emissions and advancing policies to mitigate climate change. International Climate Mitigation Agreements have been crucial in developing a conceptual framework to address the issues that contribute to the mitigation of climate impacts but also constitute another form of governance at the international level that includes the provision of budgets for the application of the agreed strategies. As such, issues of accountability are pressing. Moreover, there is a discrepancy between the slow pace of international agreements implementation and accentuated impacts of climate change when carbon emissions remain unchecked and unmitigated globally. These agreements began with sustainable practices to commitments and legal accountability. More recently, the agreements include mitigation practices, renewable energy, and adaptation.

The first international agreement effort was developed by United Nations World Commission on Environment and Development in the 1987 Brundtland Report. The term sustainable development was coined in this agreement to recognize economic growth as a leading factor in altering the environment. The Brundlandt Report sought to address environmental strategies in cooperation among members of the international community to deal more effectively with concerns to protect and enhance the environment by the year 2000. At that time the term climate change was not coined yet and instead the discourse referred to environmental issues based on the relationships between people, resources, environment, and development. The specific environmental problems addressed in the Brundtland Report were desertification, destruction of forests for agriculture, urban development, paper and wood production (among other purposes) acid precipitation, and the burning of fossil fuels to be problems that turn productive land into desserts, acidified soils and water bodies, causes of global warming, the greenhouse effect, sea level raise and coastal flooding, ozone layer depletion, ocean food chain disruption, toxic substances absorption into food and underground water (United Nations World Commission on Environment and Development 1987). World poverty and international inequality are decisive factors in environmental deterioration and according to Giddens (2011), necessary to be integrated into environmental recovery strategies. Economic development when addressing environmental issues plays a key role and needs to be considered together and not separate from each other. Sustainable global development, in this regard, can be accomplished only if population size and economic growth are congruent with the "changing productive potential of the ecosystem" (Giddens 2011, 16). It is in the Brundlandt Report that the notion of sustainability is adopted.

The Rio Earth Summit in 1990 introduced terms that all participating nations calculate their carbon emissions and report the levels annually (Meakin 1992). In the United States, the Lieberman-McCain Climate Stewardship Act of 2003 required the Administrator of the Environmental Protection Agency (EPA) to promulgate regulations to limit the greenhouse gas emissions from the electricity generation, transportation, industrial, and commercial economic sectors (as defined by EPA's Inventory of US Greenhouse Gas Emissions and Sinks) (Giddens 2011). As a way of developing a governance structure, the bill also provided for the trading of emissions allowances and reductions through a National Greenhouse Gas Database which would contain an inventory of emissions and registry of reductions. The Kyoto

Protocol (The United Nations Framework Convention on Climate Change [UNFCCC] 1998) that took place in Kyoto, Japan, in 1997 encouraged developed countries to cut their emissions by an average 5.2 percent relative to 1990 levels by the period 2008–2012. Russia ratified by the end of 2004. The Kyoto Protocol initiated the Clean Development Mechanism (CDM) allowing industrial countries to get credits to put toward their Kyoto targets by funding clean energy projects in developing states. Emissions reductions were "generally achieved, but in an erratic, highly inefficient, and corrupt manner" (Giddens 2011, 188). The 2007 Bali Action Plan or Bali Roadmap (UNFCCC 2008) designed a new negotiation process to replace Kyoto and included reducing emissions from deforestation in developing countries, development and transfer of technologies, reporting on global observing systems for climate, among other important mitigation strategies, but also reaffirming that economic and social development and poverty eradication are global priorities. Accordingly, different countries assumed the initiative to advance different technologies to aid diminishing climate change impacts: Advanced Vehicles (Canada), Bioenergy (Brazil and Italy), Carbon Capture, Use & Storage (Australia and the United Kingdom), Energy Efficiency—Buildings Sector (United States), Energy Efficiency—Industrial Sector (United States), High-Efficiency, Low-Emissions (HELE) Coal (India and Japan), Marine Energy (France), Smart Grid (Italy and Korea), Solar Energy (Germany and Spain), and Wind Energy (Denmark, Germany, and Spain), among others. The technology action plans included the mitigation potential of the technology, obstacles, and best practice strategies to overcome these obstacles, and a menu of specific potential individual and collective country actions to move to best practices and actions. These actions are designed to help reduce market barriers and realize the full potential of clean energy technologies. According to Giddens (2011) the agreement, however, did not contain concrete strategies. In 2009, the Major Economies Forum (MEF) on Energy and Climate brought together the 17 largest economies (Australia, Brazil, Canada, China, The European Union, France, Germany, India, Indonesia, Italy, Japan, Korea, Mexico, Russia, South Africa, the United Kingdom, and the United States) to facilitate dialogue and help generate political leadership for concrete initiatives and joint ventures to increase the supply of clean energy and help diminish greenhouse gas emissions.[2] According to Giddens (2011) Australia and Japan pledged to reduce their emissions by 25 percent over a 2000 base and a 1990 base, respectively, and President Obama pledged that the United States would reduce emissions by 17 percent over 2005 levels by 2020 and by more than 80 percent by 2050. The Copenhagen Accord in 2009 (UNFCCC 2009) was a major economic forum where agreement on binding carbon reductions was not achieved but it included for the first time China, India, and Brazil. One of the main goals of the Copenhagen Accord was to limit global temperature increase to 2 degrees Celsius, accountability on mitigation pledges from countries, emissions reduction, preserve forests, and adapt to climate change, and reduce deforestation, among other pressing issues. Significantly, the United States declared support for the goal of $100 billion a year for developing countries. China conceded to United States for its actions to be open to some form of international scrutiny (Pew Center on Global Climate Change 2009). In 2009 China made efforts to reduce energy intensity, promote cooperation on energy and climate change policy, and work together to achieve concrete climate mitigation results (Giddens 2011). The Cancun Agreement paralleled bottom-up framework with countries undertaking national pledges for 2020 generating political rather than legally binding actions (The Climate Institute 2010). In 2011, the Obama State of the Union Address sought to promote clean energy and clean energy technological breakthroughs (State of the Union Address, President Obama 2011). The Durban Platform for Enhanced Action—COP 17—2011 called for "a protocol, another legal instrument or an agreed outcome with legal force under the Convention applicable to all Parties, to apply from 2020, but provided no further substantive guidance" (Giddens 2011, 2, UNFCCC 2012). COP 19 in Warsaw focused on a bottom-up characteristic in an emergent agreement to request from all parties to submit intended contributions determined nationally (Giddens 2011, UNFCCC 2013).

At least once a year an agreement has been reached among different countries. The latest agreement took place in Paris in 2015 (United Nations Treaty Collection 2015). The Paris Agreement reaffirms the goal of limiting global temperature increase below 2 degrees Celsius. The agreement also commits all countries (developed and developing) to a common framework in which to regularly report on their emissions and implementation efforts under international review in nationally determined contributions every five years. The agreement also states that developed countries should support the efforts of developing countries and encourage contributions by developing countries too. The agreement also seeks to extend from 2020 to 2025 the mobilization of $100 billion a year and a higher goal after 2025 as well as to extend mechanisms to address climate change "loss and damage" without further liability or compensation resulting from those acts. It will also allow for emission reductions from one country to count for another country's nationally determined contributions (NDC). The agreement sought to provide a bottom-up flexibility along with top-down rules to promote accountability. The NDCs are to be recorded in a public registry maintained by the United Nations at least every two years to provide transparency and as a way of holding countries accountable. There is also a provision in the agreement to stock-take successive NDCs in a five-year cycle to assess collective progress toward meeting the agreement's long-term goals. These agreements constituted accords on budget allocation for the development of legally binding accountability at the international level to get as many countries to commit. The structure being created among the different agreements has evolved to produce a form of administration of economic resources toward climate change mitigation. This structure now involves carbon markets where mitigation outcomes can be internationally transferred to implement the country's NDC as tradable emission offsets. This is the development of a superstructure of trade among tradable emissions offset and as such a recent regulatory framework in the making resting on voluntary and legal enforcement.

The emphasis of these agreements focuses also on adaptation. The United Nations Framework Convention on Climate Change (UNFCCC) seeks to develop efforts to strengthen adaptation by strengthening resilience and reducing vulnerability to climate change to plan, implement, and report adaptation efforts, provide support and review of adaptation efforts in developing countries as well as evaluate adequacy and effectiveness of adaptation support. The Warsaw International Mechanism for Loss and Damage associated with Climate Change Impacts sought to work toward the global goal of adaptation mostly by "enhancing knowledge and understanding of comprehensive risk management approaches to address loss and damage associated with the adverse effects of climate change, including slow onset impacts (such as sea level rise), strengthening dialogue, coordination, coherence and synergies among relevant stakeholders, and enhancing action and support, including finance, technology and capacity-building, to address loss and damage associated with the adverse effects of climate change" (UNFCCC 2013).

In 2015 India and France established the International Solar Alliance supporting solar energy development. Twenty developed and developing countries are involved in this initiative pledging to double public investment in clean energy research and development over five years. The Breakthrough Energy Coalition is an initiative developed by Microsoft founder Bill Gates and 27 other major investors in 10 countries, pledging to double public investment in clean energy research and development over five years. The initiative recognizes the role of government in leading technological development for the protection of the common good by providing government-funded research programs for basic and applied energy research linked to private investors. The Mission Innovation: Accelerating the Clean Energy Revolution initiative has the objective of making clean energy widely affordable by reinvigorating and accelerating global clean energy innovation. Representatives from Korea, Sweden, Germany, and Australia launched at the United Nations Climate Change Conference 2015 (COP21) led this initiative in Paris 2015. As of October 5, 2017, India and the European Union strengthened their commitment to cooperate toward the clean energy transition on energy efficiency, smart grids, and low-emissions mobility. The Global Covenant of Mayors for Climate and Energy (2016) was launched by UN Secretary Ban Ki-moon and Michael Bloomberg as the special envoy for Cities and Climate Change, under the leadership of the world's global city networks and local governments for sustainability to increase the visibility as leaders responding to climate change and encouraging public and private investments in cities by meeting transparent standards, building data on the impact of city action that is consistent and robust, as well as to accelerate more ambitious, collaborative, and sustainable local climate action. At the urban level in New York City different organizations have taken the lead to provide initiatives that address climate change such as the Climate Change Adoption Task Force, 2008;[3] The New York City Panel on Climate Change 2008;[4] Green Codes Task Force 2008;[5] Mitigation Planning Council 2013;[6] Building Resiliency Task Force 2012;[7] and New York Connecticut Sustainable Communities Consortium 2012.[8] The Non-State Actor Zone for Climate Action (NAZCA) was set up under the Lima-Paris Action Agenda (UNFCCC 2015). The portal aims to accelerate cooperative climate action and support of new universal climate change agreement by registering commitments to action by companies, cities, subnational regions, and investors to address climate change. In collaboration with core data partners such as the Carbon Climate Registry, The Climate Group, the Investors on Climate Change, and the United Nations Global Compact where countries are able to record commitments and actions. These partners have an established record of data management and assessment to track commitments over time to ensure progress. They will provide data management and assessment processes as well as capabilities to track progress over time. NAZCA provides a virtual portal to collect data regarding commitments from different entities.

ENVIRONMENTAL PEACE AND COOPERATION

Dabelko (2006) advocates that through the agreements, opportunities and gains to all parties through cooperation could be achieved, despite the current levels of pollution and resource degradation. Shared environmental problems might set in motion social and political changes that could promote peace rather than violence in the issue area. The challenges that climate change events pose can generate conflict among regions but also could create opportunities to address the issue-specific technical complexity, ecological interdependence, and the need for actions to anticipate climate impacts (Dabelko 2006). Tran-societal linkages that generate new environmental responsibility norms and peaceful resolutions could result in institutional transformations that are more open, inclusive, and accountable. The kinds of conflicts generated by climate impact generally involve trans-boundary resource disputes that trigger environmentally induced insecurities. However, under these circumstances cooperation and not conflict seem to be the norm in resolving international environmental security disputes (Dabelko 2006). It is in this context that Klare (2001) points to the fact that one of the most distinctive features of environmental global security and future wars will likely be centered on the possession and control of vital economic goods and resources, water becoming one of the most critical. The regions with political and social instability where abundant resources can be found and the routes from and to these locations will likely become part of territorial conflicts. According to Klare (2001) these areas will be particularly showcased in the media and be the focus of international policy makers, where large military power will be concentrated. Under these circumstances, global cooperation rather than conflict will be most necessary to address how economic and other resources will be distributed. For Peluso and Watts (2001) violence emerges from population growth, environmental degradation, resources scarcity, and social, economic and political dislocation. Environmental scarcity, according to them, has complex consequences from extremely complex ecological and political systems since political conflict usually centers around issues of resource control. There are differentiated scarcities. For instance, demand scarcity emerges from population growth, supply-induced scarcity from environmental degradation and limits to availability of resources, and structural scarcity results from the unequal distribution of resources, politically organized. Scarcity plays out on the dissimilar impacts of population growth, environmental degradation, and inequality under the political and economic circumstances of conflict prone–affected areas (Peluso and Watts 2001).

Within the context of climate change and disaster prone areas, environmental security has been developed to address such concerns and to provide strategic support to crisis prevention through cooperation. Conflicts tend to respond to structural and contextual factors. Even though climate could trigger conflict this effect is usually accompanied by other social and political forces and in this regard, climate change generates insecurity. The New York United Nations Population Fund (2001) emphasizes the interaction between population dynamics such as size, distribution, movement, and environmental change. In this context, the reciprocal relationship between humans and environment has tended to focus on the impact of humans on environment rather than the impact of environmental change on human welfare.

Focusing on the policy component, population and environment linkages deserve some clarification points. First, the possible conflict that population and environment issues could trigger is not the only kind of relationship between humans and the environment where cooperation is also a prevalent relationship. Second, the impact of humans on the environment is unprecedented. Third, the relationship between humans and the environment concerns all political territories and countries due to the transboundary and global reach of climate change impacts. No mechanism has yet been developed nor implemented that addresses explicitly the transboundary character of climate change impacts at the national level with the exception of nation-states participating in the precarious structure of climate change agreements, shifting budget allocations and entrepreneurial capacities of urban climate change policy development. Cities, however, have increasingly become leaders in mitigation potential and city climate networks (Zomer and Hsu 2015). Increasing attention has been placed in the role of cities to face opportunities and challenges to address climate change (slow onslaught and impacts) within the characteristically slow implementation and accountability of climate change agreement international negotiations (Zomer and Hsu 2015). Paris developed the "Buffeted but not sunk: Paris Resiliency Strategy" in 2017 to address heat waves and the urban heat island effect, green infrastructure pilot program relying on collective intelligence and cooperation with other territories (Maire de Paris 2017). Among the 2000 cities involved in the Compact of Mayors that voluntarily report emission reduction targets, only 200 have plans to specifically target emission reduction and mitigation plans (Zomer and Hsu 2015). One of these cities is New York City.

PREPAREDNESS PLANS AND BUDGET ALLOCATION FOR URBAN CLIMATE CHANGE POLICIES

In New York City a number of strategic plans to mitigate climate change, prepare for, and recover from climate change events have been developed under the leadership of NYC major and NY State Governor administrations, and many of these plans were developed after the experience with Hurricane Sandy. They include PlaNYC: A Stronger, more Resilient New York (The City of New York 2013b) which after Hurricane Sandy's impact is an initiative that has been incorporated within the umbrella approach of the NYC Special Initiative for Rebuilding and Resiliency (SIRR),[9] 2013,which grouped special reports for different recovery-related issues: Multifamily Flood Insurance Affordability Study in 2016, New York State Coastal Management Program, 309 Assessment and Strategies in 2015, Building the Knowledge Base for Climate Resiliency: New York City Panel on Climate Change 2015 Report, and One City Built to Last: Transforming the New York City's Buildings for a Low-Carbon Future in 2015, One New York: The Plan for a Strong and Just City in 2015, Geothermal Systems and Their Application in New York City in 2015, and Hurricane Sandy Rebuilding Strategy: Stronger communities, A resilient Region in 2013. These plans are part of an active and bold initiative that seeks to deter the damaging effects of climate change by positioning New York City as a leading force globally. New York City is also part of the 100 Resilient City Initiative (100RC, 100resilientcities.org) under the auspices of the Rockefeller Foundation. The United Nations has also produced a 2015 Report on the Patterns of Disaster Risk Reduction Actions at Local Level. Still other plans have been developed in different affected areas around the United States, mostly after a disaster has impacted those areas. The Coastal Community Resilience Index is a tool developed for communities to determine their degree of preparedness for storms and storm recovery by the Gulf of Mexico Alliance in association with the Mississippi, Alabama Sea Grant Consortium.[10] The National Oceanic and Atmospheric Association's Coastal Storms Programs[11] help communities to build resources for resiliency, create a shared community understanding of the potential extent of future losses, apply strategies to serve near- and long-term mitigation needs, and take the first steps toward adapting to sea level rise. The target audience for this effort is Alabama, Mississippi, and Eastern Louisiana, and the initiative is supported by follow-up training and/or technical assistance. The Presidential Policy Directive / PPD-8[12] is aimed at strengthening the security and resilience of the United States through systematic preparation for the threats that pose the greatest risk to the security of the nation, including acts of terrorism, cyber-attacks, pandemics, and catastrophic natural disasters. The Presidential Policy Directive includes national preparedness systems, building and sustaining preparedness, national preparedness report, roles and responsibilities, and definitions that include national preparedness, security, resilience, prevention, protection, mitigation, response, and recovery. This directive is meant to include the responsibilities under Homeland Security for threats such as acts of terrorism, cyber-attacks, pandemics, and catastrophic natural disasters. The Presidential Policy Directive (PPD-8): National Preparedness was developed to create a shared community understanding of the potential extent of future losses, apply strategies to serve short- and long-term mitigation needs, and take the first steps toward adapting to a rise in sea level. This support will be in the form of follow-up training and/or technical assistance. As such, the "Whole Community" approach is being incorporated into all PPD-8 deliverables, including the National Preparedness Goal, National Preparedness System description, National Planning Frameworks, and the campaign to build and sustain preparedness nationwide, as well as leverage the approach in their development. In New Jersey the Department of Health provides for a Disaster Mental Health hotline.

In an attempt to add an assessment component to recovery actions implemented, Adhikari et al. (2015) study on post-Sandy resiliency planning in Hoboken and Jersey City, highlights the importance of engaging

the juncture between extreme climate events and the conditions produced by market deregulation and government retrenchment within neoliberal practices on crises mode.

Resiliency in the architecture literature generally refers to infrastructural redundancies, such as levees, that offer flexibility to structures that block flooding (Saloojee and DeBritto 2014). Basically, rather than one continuous structure, redundancies involve multiple segments with the same functionality that could be operated in a flexible capacity. I would argue that a heightened level of social flexibility is also required to deal with the uncertainty of climate change occurrence and its aftermath. Social redundancies in flexible systems of communication and organization allow for response to climate threats and ongoing recuperation measures to be in place. Redundancies that are based on flexibility are required in both infrastructure and social-community systems. Redundancy, preparedness, and adaptation are part of the generalized resiliency approach to climate change policy largely implemented by those who have placed climate change concerns at the forefront of urban development. Adaptation and mitigation global approach and local interventions within the resiliency discourse should not be confused. Climate change mitigation or climate protection represents a local effort to solve the intrinsically global problem of an interconnected ecological system in order to protect what could be construed as a global common good. Adaptation to climate change involves local consequences of that global phenomenon. As such these consequences are experienced locally and the practices set forth for adaptation produce not a global but a local common good even when the best practices for adaptation, preparedness, and mitigation can inspire implementation practices in other locales.[1]

RESILIENCY

According to Saloojee and DeBritto (2014) resiliency was originally used as a measure of materials strength and it is still used in mechanical engineering. The modulus of resilience defined the force required to rupture a material and therefore test materials resistance to extreme conditions. The term is still applied to design codes for structural, civil, and mechanical engineers, and naval architects. Resilience in objects involves the capacity to retain certain characteristics and continue performing within the parameters of disturbance (Saloojee and DeBritto 2014). Resilience therefore implies a dynamics of identity to address the conditions under which a thing remains with its distinct qualities in spite of change. Modern conceptualizations of resilience emphasize an ecological approach pushed forward by C. S. Holling which still influences resiliency thinking, strategies, and concepts in economics, security disaster planning, psychology, and ecology (1973). Holling coined the term in 1973 to signify an ecosystem's ability to change while prevailing. Ecological resilience refers to an ecosystem that remains stable despite disturbances in a fast recovery to a stable state. In line with this approach, resiliency then is the ability for an ecosystem to adapt quickly after been deeply affected by external factors instead of remaining consistent in the face of dramatic change (1973). Within this framework, adaptability is considered more important than the immediacy and consistency of the system's recuperation. There is an underlying notion of system stability that has been recently questioned and instead considers ecosystems to rather be in constant flux than in an assumed stable state. In that regard Klein et al. (2004) define resiliency as "the self-organizing capacity to preserve actual and potential functions under constantly changing circumstances" (40). Resiliency, according to Saloojee and DeBritto (2014), characterizes a system that is dynamic. For them a resilient infrastructure is one where the system is self-organizing but with limits to this self-organizing capacity, which occurs in adaptive cycles and multiple scales. The system is interconnected to social, economic, and biophysical realms. Resilience helps specify general and transformability dimensions. Accordingly, working with resilience includes processes of adaptation and transformation, something is gained and something is lost. Resiliency does not preclude knowing everything and it is not about not changing. In other words, resilience is a process that involves the ability to absorb, adapt, and learn in the face of change.

The discourse on resiliency is a socio-ecological approach that has emerged out of the uncertain character of climate change due to carbon emissions that along with the production of energy based on fossil fuels and rapid urbanization have deteriorated natural ecosystems. Resiliency addresses different aspects of the adaptive characteristics required to implement climate change protocols. According to Giddens (2011), resilience is defined as adaptive capacity "to cope but also respond actively and positively to external changes or shocks" (164). If it is a change to the physical environment that increases vulnerability to flooding, strengthening, repairing, or building new systems in the infrastructure might be considered resilient strategies. Regarding individuals and/or groups affected by a dramatic change, the capacity of its members to act in unison rather than divided and fragmented, and the ability to modify and transform when necessary, also characterizes resiliency. Extreme climate events such as hurricanes, droughts, floods, bush fires, heat waves, tornadoes, tropical storms, typhoons, winter storms, hailstorms, dust storms, and waves of cold weather are some of the strong reminders of the power that climatic fluctuations have over people's lives and to which people need to respond in a resilient manner. The increase in these climate events has made of resiliency a focus to cities' climate change mitigation plans.

Gupta et al. (2016) define resilience in their Case Studies of Community Resiliency Policy Report for the National Institute of Standards and Technology, US Department of Congress, from a socio-ecological perspective as "a system's capacity to absorb disturbance and re-organize into a fully functioning system" (6). Their work focuses on communities' ability to adapt to changing conditions as well as resistance and recovery from emergency disruptions due to naturally occurring threats, accidents, or deliberate attacks. Resiliency therefore is understood as a trajectory of positive adaptive–linked capacities after a disturbance that allows coping mechanisms with minimum impact and damage. In this context the outcome or goal of a resilient approach, losses are avoided or reduced, the hazard is endured, and recovery involves minimal social disruptions. In the New York State 2100 Commission Report (2013) produced after Sandy and sponsored by New York Governor Cuomo, resiliency is defined as redundancy in the infrastructure. This redundancy involves a way of thinking about systems in which different parts perform similar functions, in a repetitive manner, providing flexibility to the system by not relying on just one centralized functionality. The discourse on resiliency that has emerged post Sandy includes not only infrastructural components of the physical environment but also ecological and demographic shifts. Social aspects of the areas impacted by disasters also figure as relevant and in need of serious inclusion in the preparation and recovery process. According to the New York State 2100 Commission Report (2013), infrastructure redundancy should provide continuity of services to communities and the ability to rebound in the aftermath of an extreme event. A continued learning effort is necessary to improve resilient practices outside of extreme event impacts.

Plan NYC (The City of New York, Mayor Michael Bloomberg 2013b) states that a resilient city is "protected by effective defenses and adapted to mitigate most climate impacts; and second, able to bounce back more quickly when those defenses are breached from time to time" (7). Accordingly, the plan is based on the principles of protecting the coastline and waterfront infrastructure as part of the natural and cultural resources of the city. The plan also proposes to invest smart and effectively for the protection of the city in an experimental approach to modify and expand strategies as well as pilot projects that could be implemented at a large scale over time to all New Yorkers and not only those dramatically affected by Hurricane Sandy, heat waves, or power outages. The plan emerged as a wakeup call after Hurricane Sandy's impact in an opportunity to hone the learning capabilities that catastrophes and disasters afford. For Tompkins and Adger (2004) resiliency is an important sustainable response to climate change that involves a system that is able to cope with extremes within gradual changes in mean climate parameters and "absorb perturbations without being undermined or becoming unable to adapt and learn" (14). Adaptive management processes are informed by iterative learning of earlier successes and failures taking into account the complexity of the ecosystem. Resiliency then creates a space that provides the opportunity for reflexivity, to step back and think about how to move forward. The resilience of systems is defined by their ability to self-organize and two important aspects to the ability of a system to adapt are "robustness to uncertainty and flexibility or ability to change in response to altered circumstances" (Adger et al. 2005, 81). Adaptation therefore is a key component to a robust approach to resiliency. Berkes and Jolly (2001) identified three main characteristics of change under the organizing concept of resilience. One characteristic is that under resilience a system can go through major changes while remaining with control over function and structure is required. The second characteristic includes the degree to which the system is capable to self-organize, and the third characteristic is that of the community to be able to build and increase its capacity for learning and adaptation (Berkes and Jolly 2001).

ADAPTATION

Self-organization, learning, and adaptation are key concepts for conversations about climate change impacts, as defined by the Resilience Alliance.[2] Although the concept of resilience is most commonly used in the study of ecosystem dynamics (Holling 1973), it can also be applied to social systems (Adger 2000; Adger et al. 2001), social-ecological systems (Gunderson et al. 2010; Berkes and Folke 1998; Gunderson and Holling 2002), and the study of global climate change (Holling 1997). According to Adger et al. (2005), resiliency involves the ability to adapt to challenging circumstances. Vulnerability to climate change is largely intensified within marginalized communities who cannot bounce back from a climate event impact as quickly and resourcefully which accentuates the challenges. Their dependency on the government to alleviate recovery costs lengthens the amount of recovery time after impacts from changing environmental conditions. Therefore, there is a need to develop a broader conceptualization of what would constitute a more effective, legitimate, and equitable response to resiliency efforts.

Adaptation, however, is not about returning to some prior state. The modernist ideology about a linear trajectory of an ontology of development based on progressive increments does not quite fit expectations on recovery after a climate event when the complex landscape of factors involved and the dissimilar conditions under which development happens are not acknowledged. The assumption of a progressive linear sequence of events to development denies the multiple possible outcomes of life as a constantly changing process. Social and natural systems coevolve with each other over time, changing constantly, and a climate event involves an accelerated and intensified change. Coevolution could be understood as the ability to accept the immanence of change. The expectations that might unfold after a climate event could be unrealistic if the positivist linear progression of development is assumed and remains unchallenged. Things might turn worst or better than before the climate event. Uncertainty in the face of risk introduces a key component to considerations about adaptation. As conveyed in an informal interview with a landscape architect who designs for flood prone areas, unpredictability and change, dealing with the unknown, and the possibility of tentative solutions are required skills to deal with the challenges posed by climate change. Learning from the process of recuperation as trial and error, and as alternative options to be considered for implementation in an experimental and exploratory fashion, is broadly discussed in the literature on adaptation to climate change, and to climate events more specifically (Adger et al. 2004; Berkes and Jolly 2001; Grothmann and Patt 2005).

Grothman and Patt (2005) address the process of adaptation to climate change in four stages: signal detection, evaluation, decision and response, and feedback. Their consideration of adaptation is defined in recurring steps that include collecting information and building awareness, planning and design, and monitoring and implementation. But adaptation and resiliency need to be placed into larger contexts for a full understanding of the factors that affect preparedness and response to climate change. Gupta et al. (2016) state that adaptation involves the actions taken in response or in anticipation to projected or actual changes in climate. Accordingly, the main goal of adaptation is to minimize adverse impacts but also and often undermined is the ability to take advantage of the opportunities posed by climate change in terms of an unprecedented and uncertain situation. These can be addressed with curiosity in a trial and error approach to improve physical infrastructure design as well as communication among key stakeholders and affected constituencies. Tompkins and Adger (2004) define adaptation to climate change as adjusting to "ecological, social or economic systems in response to observed or expected changes in climatic stimuli and their effects and impacts in order to alleviate adverse impacts of change or take advantage of new opportunities" (77). Future "surprises" regarding climate impacts or the unknowable risks posed by these require adaptation by building on the ability of individuals, groups, and/or organizations to implement decisions and actions. In that line of thought, adaptation is then a constant set of activities, decisions, actions, and attitudes about the different aspects of life that have consequences for social norms and processes. Adaptation options could be grouped under their purpose or objective, mode of implementation, and/or the institutional form they take. Adaptation occurs within the context of cultural, demographic, and economic change, therefore information technologies, global governance, social conventions, and the globalizing flows of capital and labor all influence adaptive strategies. In light of this, climate change adaptation decisions are difficult to extricate from decisions and actions generated by other social and economic situations and circumstances (Gupta et al. 2016). Adaptation to climate change is a complex process that includes short- and long-term strategies and costs considerations. Adaptation could also lead to unsustainable practices such as an increase demand for air conditioner during a heat wave or higher property values for low-lying ocean front properties. The perils of adaptation or when adaptation backfires usually happen when adaptation implies a preservation of the status quo or the preservation of how things were rather than the acceptance of the change that have transformed individuals and communities. A conservative approach to adaptation could also be assumed if the maintenance of the status quo means that those who are most affected and with less resources have to accept a transformation that worsens their opportunities to bounce back to an acceptable standard of living. Adaptation for them could mean struggling with governmental agencies to become aware of aid programs available, lack of or inadequate insurance, contractor's fraud and the challenges of moving through the thick bureaucracy in order to get aid. For them, adaptation and resilience could mean being caught in a maze of uncertainty and red tape.

In building adaptive capacity, communication of climate change information is key as well as developing awareness about potential impacts, preserving well-being, protecting property and land, maintaining economic growth, and seizing new opportunities (Gupta et al. 2016). A key challenge to purposeful adaptation implementation and effective adaptation practice rests on the integration of adaptation actions and policy across different sectors. Measures to ensure specific populations to recover from loss include increased access to resources and insurance, increased reservoir of storage capacity, vegetation buffers to withstand climate variability, and flood-control on ground floors in new building construction, to name a few (Gupta et al. 2016). Hazard preparedness and climate change mitigation activities can alter the exposure to the effects of climate change. These climate adaptation strategies vary as does the role of international and national policy, and individual and collective action. Collective investment is also an adaptive strategy, such as the case of coastal communities that galvanize support for sea defenses according to regulatory frameworks. Gupta et al. identify legitimacy in the process of developing adaptive strategies as an important aspect to evaluate how acceptable decisions are made and the underlying distribution of power that affect participants and nonparticipants in the decision-making process (2016). Legitimacy and compromise through the evolution of adaptation strategies in the experimental trial and error process highlight the effectiveness of most climate change strategies. In lack of universal rules that could warrant legitimacy in policy implementations, cultural interpretations and expectations ultimately determine what is considered legitimate or not. Equity and fair public action as granted by equitable and legitimate practices is a crucial aspect for long-term sustainability (Gupta et al. 2016). The level of integration and/or exclusion that different populations experience affects their access to resources, including successful adaptation strategies. Gupta et al.'s description of the economic growth conditions and the kind of adaptation strategies implemented and/or pursued show that the more homogeneous per capita income, larger population, and a heterogeneous market–led approach in a specific society, along with a strong focus on economic efficiency and welfare maximization at an aggregate level, there tends to be less concern about welfare loss or increased vulnerability for marginalized sectors (2016). The adaptation policy for this kind of society tends to evaluate effectiveness based on infrastructure performance while little attention is paid to the equity or legitimacy in the adaptation strategies process.[3] Adaptive strategies that consider global solutions to achieve economic, social, and environmental sustainability based on a global scale of cooperation and regulations to introduce clean and efficient technologies are also part of the continuum to determine the kind of society fit with adaptive strategies (Gupta et al. 2016). The last stage on this continuum in societies with low rate of population growth and environmental, economic, and socially sustainable, locally oriented development, is to identify economic efficiency as relevant but focused on the distributional effects within local regions. A strong emphasis on equity and legitimacy promotes a social contract for sustainability and explicit consent for adaptation actions at multiple levels of the government. Gupta et al. (2016) conclude that adaptation strategies considering markets in insurance, coastal planning, health interventions, built environment, water resources, and adjustments and adaptations within resource-based livelihood will receive investments that are episodic and staggered, and after a large-scale impact would be more needed. Considerations about adaptation strategies and their implementation would likely be triggered by extreme events rather than slow-onslaught. There is therefore the importance to increase awareness to climate change at the policy-making level to develop government legitimacy. For instance, Miller and Rivera (2008) argue that disasters shed light on previously unaddressed political issues and a disaster, in this case Hurricane Katrina, became an accelerating agent for policy changes. A similar outcome was experienced after Hurricane Sandy in New York City.

Adaptive management processes involved in climate change across multiple scales need to be informed by a dynamic learning process about the ecosystem including natural and built elements. Successes and failures previously faced also need to be assessed in order to increase current resiliency to better improve the ability to respond and adapt to long-term climate change threats (Adger et al. 2005). In developing such a type of adaptive management to climate change threats, a larger degree of flexibility in the institutions carrying these programs for resource management are necessary. Accordingly, sustainable development needs to be understood broadly to include equality in the opportunity to adapt. The governance issues associated with adaptive management require long-term demographic and consumption patterns transition, large-scale infrastructure developed for adaptation, new technologies in agriculture and natural resource use, as well as plans and policies that help conserve natural areas and ecosystems (Adger et al. 2005). Conservation in this context needs to be inclusive of people as a social component, rather than only limiting considerations of ecosystems by endorsing social acceptance of shared resilience goals (Adger et al. 2005).

In urban areas governance at the community level, participatory structures most strive to effectively deal with social exclusion and the urban underclass. When these communities are/or become tolerant and mixed along diverse ethnic, social class, race, sexual preferences and abilities, the process toward adaptation to climate change is facilitated (Tompkins and Adger 2004). This was increasingly important in the Rockaways where Senior Living Housing and Housing for people with Disabilities became most challenging when the power outage in high rises trapped people inside for weeks. Volunteers needed to go into these buildings to check for the elderly and the disabled who could not travel outside their homes. Tompkins and Adger (2004) also identified the needs to recognize indigenous knowledge and technologies, at times more in tune with natural cycles and solutions than our advanced urban societies. Flexibility within cultural traditions and networks provide alternatives such as food sharing and intercommunity trade for needed resources (Berkes and Jolly 2001). The nature of the relationships between community members is also critical, as are access to and participation in wider decision-making processes. A sense of powerlessness after a disaster might be due to lack of access to decision makers, among other factors. Networks of engagement are evaluated also in terms of the support they offer to the dependence networks of participants in vulnerable positions. Sharing information and feedback is most important to keep the channels of communication open and effective. As with preparedness, information dissemination and raising awareness about adaptive practices has to be shared. Biesbrock et al. (2010) examined Europe's adaptation capability to climate change and recognized that a climate event is a motivating factor to take action. Other factors that facilitate response to an extreme climate event involve "political will, human and financial resources, good co-ordination between key actors in different sectors and at different administrative levels, and compatibility with other policies" (Biesbrock et al. 2010, 15).

Accordingly, Gonzalez et al. (2016) portend that adapting to the circumstances triggered by climate change impacts usually involves structural measures such as dams and levees as well as nonstructural measures such as elevation of property and buyouts, and pre-disaster planning. Increasingly federal agencies, local, and state governments are more inclined to consider and develop implementation projects in nonstructural measures and strategies that involve vulnerable communities rather than the usual almost automatic response to target the physical infrastructure. Nonstructural strategies are flexible, multilevel operation from individuals to communities, and seek to offer additional protection in vulnerable areas. The nonstructural adaptive strategies could offer long-term and permanent protection or reduce the risk to people and property against natural hazards, and/or provide protection to areas excluded from hard infrastructure protections. Usually, both structural and nonstructural strategies are needed. For Gonzalez et al. (2016) individuals' and communities' effective adaptation and increase in resilience from the effects of climate change and their ability to cope with climate change are enhanced with nonstructural adaptation.

The 2012 Coastal Master Plan in Louisiana is an example of the importance of structural programs as it identifies alternatives to reduce flood risk by improving design and elevation of residences. Structural adaptation strategies provide protection and risk reduction to Gulf Coast communities against sea-level rise, changes in precipitation patterns and sea surface temperature changes that could increase the frequency of intense hurricanes. However, these adaptive strategies need to include socially vulnerable communities that might not have the financial resources, knowledge, legal, technical, and social support required to secure the funding to reduce their overall risk.[4]

In dealing with the changes prompted by extreme climate events, Miller and Rivera (2008) describe the challenge New Orleans residents faced with Hurricane Katrina, to point out the trauma caused by the disruption between people and their environments. Disasters of this degree trigger collective stress by the cultural change involved and the stress of coping with a now foreign situation and reality disjuncture of no shared group assumptions. This kind of structural change disrupts the routines and social networks of the community. The traumatic situation if compounded by the sense of uncertainty, loss of control, alienation, and the issues carried along by the threat to the belief system that an extreme climate event poses for those affected (Ritchie and Gill 2007). According to Miller and Rivera (2008) for "many survivors who entered the city in the immediate aftermath of the devastation, the landscape appeared foreign to them" (112). The vulnerability that communities face in the aftermath of an extreme climate event increases in intensity by historically produced patterns of vulnerability palpable in the location, conditions of the infrastructure, sociopolitical organization, production and distribution networks, and the ideology of the society. The pattern of vulnerability in a society is replicated and often accentuated by an extreme climate event.

Giddens (2011) also indicated that preparedness as efforts to limit the warming of the world's climate require for adaptation to be anticipatory and preventative. Adaptation strategies therefore need to be organized before the climate event and prepared for possible futures; in other words, proactive adaptation needs to be a priority. This is accomplished by diagnosing and responding to vulnerabilities and the risk of confronting damage to activities, lifestyle, or resources that are valued by the community affected. In this regard, in order for the governments to promote adaptation, mitigation strategies toward the decrease in the warming of the climate must support innovation and creativity in the development of strategies from both the business sector and civil society. Adaptation and mitigation practices could be implemented simultaneously such as in the case of supporting the efficient use of water, better weather prediction systems, and/or crops that could withstand severe weather conditions. Thinking ahead in a systematic way involves planning as an adaptive strategy. Insurance practices need to also be considered within an adaptation framework (Giddens 2011).

Resiliency is based on adaptation strategies. Adhikari et al. (2015) analyzed post-Sandy resiliency planning in Hoboken and Jersey City and found that the Disaster Risk Reduction (DRR) actions at the local level were insufficient. Even after a large-scale natural disaster like Sandy, "local governments do not automatically embrace the various mechanisms of preparedness, mitigation and resiliency and forums, studies, and panels on disaster and resiliency organized by federal structures have not significantly been implemented into the integration with broad local level policy" (Adhikari et al. 2015, 8). In addition, Adhikari et al. state that the DRR seem to implement a top-down approach where demands by civic groups and local-level NGOs are not part of the negotiated process (2015). Political elites seem to be driven by a different set of available information directing decisions on the disaster preparedness and mitigation policies adopted. It is in this regard that the emphasis on community addressing climate change adaptation strategies is most relevant.

COMMUNITY RESILIENCE

Adger et al. (2005) emphasize how some natural and social systems have to learn how to become resilient through networks and institutions that can help facilitate the ability to build resilience in both social and ecological systems. Accordingly, actions to adapt and maintain resilience require adjustments to the inbuilt abilities to bounce back from challenging circumstances, at the different levels of individuals, governments, and communities. Adaptive capacity depends on how resources are distributed among individuals and institutions. The necessary multi-scalar approach to climate change response to the different ways that climate change affects communities would likely require adequate adaptation responses on multiple levels. Since climate change impacts are exposed in particular localities differently, the institutional scale for adaptation strategies needs to seriously integrate institutional resource management at the local or municipal level. The necessary interactions across scales are limited or enhanced by the power dynamics in the relationships among different institutional constituencies (Adger et al. 2005).

Nayanee et al. (2016) depart from a governance perspective regarding disaster events preparedness, response, and recovery to point out that the administration of aid at the state and community scale is based on decisions about the support galvanized from the national level in terms of federal policies, programs, and initiatives. Resiliency planning at the community level centers around trade-offs that are tangible decisions on resource allocation for disaster preparation, risk awareness, and recovery plans. It is part of the resiliency planning of the National Institute of Standards and Technology (NIST) Community Disaster Resilience Program to convene community stakeholders, subject matter experts, practitioners on disaster preparedness within a multifaceted approach to develop strategies that support the resilience of communities and aid decision makers (Nayanee et al. 2016). The identification of the social needs of the community is an integral component to understand the recovery from disruptive events by the types of services that the built environment should provide. The long-term community planning process from municipal or regional governments provided by the NIST's "Community Resilience Planning Guide for Buildings and Infrastructure Systems" helps determine, develop plans, and identify approaches and actions to improve built environment performance and community resilience.[5]

Cutler et al. (2008) identify the limitations of vulnerability models commonly used that fail to address the root causes of social and temporal dimensions of vulnerability and resilience by not including social factors at the local level and excluding vulnerability from the natural environment (Cutler et al. 2008). In an attempt to overcome such limitations Cutler et al. developed the Disaster Resilience of Place (DROP) model, which is a framework for measuring disaster resilience at the local and community levels by characterizing the relationship between vulnerability and resilience (2008). The DROP model was considered a starting point for developing a conceptual framework to address the complexity of factors linked to extreme situations in terms of social and infrastructural resiliency. The model considers rapid-onset hazards (natural or otherwise) that demand specific pre- and post-event preparation and response to say, an extreme climate event.[6] The social resilience of places is also a focus in the DROP model. Thus, indicators of social resilience include demographic characteristics and access to resources along with social resilience at the community level which differs from other models that assess resilience at the regional and global levels but do not include the community (Cutler et al. 2008).

Resiliency also includes psychological resilience or the individual's ability to adapt to stress and adversity. Individual and social resiliencies are, in addition to infrastructural redundancies, basic components to climate change adaptive strategies in urban ecological systems. Community redundancies need to exist in a culture of resilience that facilitates cooperation rather than competition for resources to offer tentative solutions. Self-efficacy, a sense of community, and a better community capacity to respond to sudden changes conform adaptive strategies. Inclusion of those who are mostly affected by the climatic change could prevent an overemphasis on self-interest and provide fertile ground for the development of more collaborative strategies. Communities of information redundancy and support are forms of implementing adaptive practices toward resiliency. In the DROP model, community resilience is based on a framework that interconnects different adaptive capacities. Community capacity is defined as the characteristics that affect the ability to identify, mobilize, and address social and public health problems through the utilization of knowledge, skills, and resources that are applicable to different circumstances and by matching changes at the individual level with public health goals and objectives. In most resilience networks the social dimension is a highly neglected aspect. In the DROP model the social dimension indicates the importance of not only demographics such as race, age, class, gender, and occupation and the social networks existing in a community but also the quality of social embeddedness in those networks, community values and cohesion, and faith-based organizations who tend to assume a crucial role in disaster recovery (Cutler et al. 2008). The DROP model provides useful measures to integrate the social dimension to resiliency.

Community resilience depends on citizen participation and the functional needs of the community defined by community engagement in formal organizations such as religious groups and electoral processes. Research indicates that three interconnected areas of citizen participation—reasons to participate and why, how the organizations and other groups affect participation, and how participation affects the conditions of the community—help explain the nature and extent of bottom-up participation in hazard mitigation and resilience efforts (Norris et al. 2008). Consequently, the social dimensions of this framework also attempt to account for the vulnerable population of the community (i.e., those who most need post-disaster support and services but might have the most difficulty accessing them). For Jacob Remes the existence of egalitarianism, inclusivity, high social capital, dense social networks, and democratic political cultures constitute the best possible community

(Perry 2017). Lack of any of these factors would translate into higher community vulnerability in light of climate change impacts.

The economic development dimension of community resilience includes stability of livelihoods, equitable distribution of income and assets in the population and economic growth. Related to community resilience, the economic dimension component to resilience is based on the notion that community resilience is dependent on the volume of economic resources available to a community as well as the resource diversity and distribution (Norris et al. 2008). Over-dependence on a narrow number of natural resources is likely to augment the variation on incomes and lower social resilience. Moreover, poor communities usually have a greater risk to be affected by the damage caused by a disaster and also encounter obstacles to mobilize support and recovery from disasters. To the extent that poor communities face more challenges to recovery from a climate change event, the population therefore becomes more vulnerable to change (Kelly and Adger 2000). Berkes and Jolly (2001) studied adaptation and change as part of a social-ecological resilience approach in a Canadian Western Artic small community of Inuvialuit people who have been tackling climate change throughout the 1990s. Adaptation to climate change according to Berkes and Jolly (2001) involves multiple scales. Long-term adaptive strategies relating to cultural and ecological adaptation in a highly variable and uncertain environment include flexibility in seasonal hunting patterns, preservation of traditional knowledge that allows the community diverse hunting activities, networks for sharing food and other resources, and intercommunity trade. Feedback from individuals, households, and the community in general is obtained regarding their responses to climate change. The development of co-management institutions also helps facilitate communication across different levels to provide feedback, learn, and self-organize and in that way communicate local concerns to regional, national, and international levels (Berkes and Jolly 2011). Even though this case study takes place in a remote Artic village and not an urban environment, it highlights the importance of involvement and emphasizes communication and feedback that is necessary at the local community. Likewise, Corfee-Morlot et al. (2009) highlight the importance of information dissemination to promote information sharing and cross-scale learning.

Amid market-disciplinary urban policy initiatives, the conditions for promoting and maintaining economic competitiveness have been reconceptualized by many local political economic elites to include diverse administrative, social, and ecological criteria (Clarke 2009, 64). A reactive response to neoliberalism and its entrepreneurial tendencies, confounded by austerity measures in the wake of climate change events, leads to a more effective organization via groups, private individuals, and/or government actors who are critical of neoliberal spotty responses. The ability to build reliable financing for action is critical (Corfee-Morlot et al. 2009). A likely accidental but necessary outcome from an extreme climate event within the context of entrepreneurial neoliberalism could also provide fertile grounds for spontaneous, bottom-up, new kind of commons where alternatives emerge to offer solutions and fill in the gap left by a free-market economy and a market enabling but social policy retrenched government. Rollout neoliberalism might provide openings for more progressive, radical democratic re-appropriations of city space that coexist within the market-disciplinary neoliberal agenda entrenched within the underlying institutional structures of urban governance (Clarke 2009). According to Clive (2005), the residualization of social life as a result of the neoliberalization process fails to account for new forms of "individualized collective action" (Micheletti 2003). Organization initiatives, alternatives to the neoliberal state after Sandy will be explored in the following chapter.

NOTES

1. Thanks to Jorg Kemmerzel for his insightful observation regarding the distinction between adaptation and mitigation practices, global issues and local implementations.
2. www.resalliance.org.
3. These tendencies run along a continuum where more or less market led, population size, economic growth, and diversity in income and technological change are considered.
4. Many of the nonphysical, nonstructural adaptation strategies center around identifying areas of risk, pre-disaster planning, post-disaster recovery strategies, floodplain regulations and flood insurance implementation, land use regulations or zoning ordinances implementation, and public education on the risk from natural hazards. Nonstructural adaptation strategies include compensation relocation to less vulnerable areas for homeowners (Gonzalez et al. 2016).
5. The National Institute of Standards and Technology (NIST) Community Disaster Resilience Program NIST is a six-step guide to hazard mitigation, risk management, and response and recovery efforts within a traditional framework. The NIST guide takes into account organizational influences; expectations about service and performance; organizational, social, planning, and technical aspects of community resilience planning and implementation customized to each community distinct challenges (Adhikari et al. 2015).
6. Slower time horizons for events like global temperature variations, sea-level rise, or drought require a different set of considerations to the adaptive capacity of the community in addition to resilience.

Chapter 4

Alternatives

Social and Physical Redundancies

Following Hurricane Sandy's impact and likely influenced by New York City's importance as an urban center, many alternatives emerged. Most alternatives devised under resiliency strategies for adaptation in flood prone areas involve both redundancies in physical and social structures. This chapter will examine the social alternatives at the community level that are crucial to effective disaster management, before and after a climate event. Occupy Sandy bottom-up organization provided the needed ongoing aid to recovery. Traditional forms of governance led by government agencies post-Sandy recovery were lacking on ongoing and efficient strategies largely due to the scale of the event in terms of preparedness but also the top-down governmental organizational approach. Therefore, the role of the community as social capital in the recovery process has been identified in the literature as a key component to successful resilience strategies for adaptation and preparedness (Miller and Rivera 2015; Dimmer and Lindberg 2014; Blachman-Biach et al. 2013; Berkes and Jolly 2001; Lin 2001). It became abundantly clear that resiliency in terms of physical infrastructure and social in terms of community involvement are the two most important components for recovery plans after Superstorm Sandy's impact. The emphasis on the social resiliency of communities is also an alternative to traditional forms of governance in the delivery of aid to affected communities. The governmental response proved insufficient for an event of the magnitude of Superstorm Sandy. This chapter will address first community redundancy, Federal Emergency Management Agency's (FEMA) "Whole Community" approach, and Occupy Sandy to account for the social-community process involved in the delivery of aid immediately after Hurricane Sandy's impact. Design alternatives to the infrastructure in flood prone areas including interventions in the ecology of coastal areas will be addressed last.

Figure 4.1 The Reconstructed Rockaway Boardwalk in 2017 after being Destroyed by Hurricane Sandy in 2012, New York City. Photo by J. Nevárez.

Superstorm Sandy made evident the lack of preparedness to extreme climate events in New York City. The alternatives inspired by Sandy have been multiple. An Occupy Wall Street working group, "Occupy Sandy" along with other groups became an efficient working example on how to deal with a disaster like Sandy by developing a community of redundancy and a culture of resiliency by offering locally coordinated support in place via relief hubs. The challenge to deal with dramatic change after a climate event impact is best interpreted from an informal conversation that took place on April 23, 2014, with a landscape architect who became involved with post-Sandy recovery:

> How communities can continue to exist and thrive where they are . . . if a community is here along the water . . . it is not the same for every situation . . . it has to be about some ability to sustain some type of use there or activity while at the same time understanding that that use or activity sustained there might change, so a community or a place might still exist, might still be there and be a very wholesome great thing but it might actually be quite different . . . so the term resilience . . . if my name is Bob and live here two blocks by the water, I am invested in this community, here doing a fulfilling thing for me in 80 years, my house might still be here but it might be further down or lifted . . . or it might be that all the houses might be moved way up or sitting on something else . . .

> accepting the eminency of change but at the same time embracing that in a way that doesn't mean that your life has to be something that you would not recognize or something completely different, you can still live there, you can still go to your favorite bar, industry can still happen, etc. they might just be different.

At a more personal level, the challenges posed by Hurricane Sandy to the residents of different coastal areas are best described in this excerpt from an email communication that was received on April 29, 2014, with a resident and architect from Long Island:

> My home was flooded during Sandy. We chose not to evacuate because of false reports regarding Hurricane Irene the previous year. We learned our lesson after going through Sandy. We were proactive right after the disaster, calling FEMA, applying for assistance immediately. That proved to help in the long run. Being knowledgeable before and after a storm really makes a difference in how you bounce back. Many local nonprofits, friends, family, and complete strangers came to our doorstep to offer their help which was something amazing to see. Our entire town is changing. Homes are being lifted above the existing ground plane anywhere from 4-12 feet in the air. Every week a new home is being lifted. Some stay in the air on cribbing for weeks, while others go up and come

Figure 4.2 A Road with a Bike Lane in the Rockaways, New York, 2017. Photo by J. Nevárez.

> down fairly quickly. Some homeowners pay for the lifts out of pocket, and some have gotten government aid funding (which is obviously much slower.) The way my town and every other adjacent town looks will never be the same again.

The National Guard, the Army, and the police assumed a distinct and powerful presence in the territories affected by Hurricane Sandy, especially in the Rockaways. Even though possibly necessary and a generalized response, deploying the military embeds a lack of flexibility to a crises that a climate event poses. During Sandy, governmental response was available but in many instances inefficient and delayed, pointing at the need to rethink and re-strategize disaster response. Generally speaking, there is an overdependence on the government to assume leadership and deliver aid during the recovery from a crisis, especially climatic ones. Characteristic of this governmental response is the deployment of the military.

ALTERNATIVE LOCALIZED RESPONSE: OCCUPY SANDY

Communities of support did emerge such as the coordinated efforts by different volunteer groups that concentrated different services in the same place as a relief hub in Beach 95 in the Rockaways. Time's Up, an urban environmental group that seeks to diminish carbon emissions by advocating for bicycle use in the city, organized rides with volunteers to carry donated items to the Rockaways. They also brought along bikes that provided sources of clean energy to recharge cell phones and other small electronics. They largely relied on the coordination of a working group that emerged from the Occupy Wall Street movement at Zucotti Park: Occupy Sandy. Occupy Sandy and other organizations collected donations at the Church of St. Luke and St. Mathew in Clinton Hill. From there cyclists would transport urgently needed items requested by residents from the Rockaways communities. In the Rockaways at Beach 95, a house not badly damaged by Sandy and owned by a nearby restaurant was made available as a storage depot to receive all donated goods. There were rooms assigned to different products: water, batteries, food, cables, lights and electric wires, first-aid kits, personal hygiene products, and other goods. Calls were placed to the house depot and volunteers would gather the requested items, mostly cleaning and hygiene items. Cyclists would deliver the requested goods to the specific houses in the Rockaways' nearby neighborhoods. In front of the storage depot house, two large solar panels that were also donated were installed. Next to the house a food truck provided food, and hot and cold beverages. Cooked meals were constantly arriving. Picnic tables were set in this area where donated food arrived on an ongoing basis. At one point a restaurant donated food cooked on the premises in a barbecue including chicken, ribs, and pulled pork for everyone free of charge. In the adjacent lot, next to the food truck and the picnic table, tarps housed donated clothes available for those who might needed them. Across the street from the house, an impromptu distribution storefront provided immediate assistance to requests for all kinds of items that people would ask for at the window. At the corner diagonally across from the distribution storefront, bicycles that generate clean energy by pedaling, brought by Time's Up were located to recharge cell phones and other small electronics. At a table next to the bicycles legal aid lawyers offered pro bono legal advice mainly for residents dealing with absentee landlords. All the goods distributed and services provided were free of charge. This area became a disaster response community of support, kindness, and camaraderie despite the tragic landscape of destruction that surrounded it. Red Cross assistance was present but transitory: a school bus with supplies parked on the street until the supplies lasted. Similarly, the Clinton Foundation used the parking lot of a shopping mall to deliver supplies also while they lasted. The National Guard and the police were present throughout the area, directing traffic since there was no electric power and the streetlights were not functioning in the Rockaways, like in many coastal communities in New York and New Jersey. The coordination of services and volunteers exemplified a community of solidarity that understood the pressing needs of the areas affected and coordinated efforts organically among different organizations and volunteers. The effectiveness of this approach caught the attention of Homeland Security and a study was developed to better understand the Occupy Sandy approach to recovery.

HOMELAND SECURITY AND OCCUPY SANDY

Homeland Security oversees all the events that pose risks and vulnerabilities that affect national security; these include both man-made/terrorist attacks and natural disasters. Homeland Security's approach toward man-made disasters such as terrorism might not be the best and most appropriate response to a natural disaster. Their coordination of efforts tend to rely mainly on law enforcement and to coordinate disaster relief with FEMA. Their heavily militarized approach to natural emergencies might be limiting and challenging to accommodate social and community concerns, broadly understood. Their role emphasizes surveillance and the prevention of looting and criminal acts rather than the ongoing support to basic everyday needs disrupted in the community. In the context of Hurricane Sandy they were able to identify the effective model implemented by Occupy Sandy. Blachman-Biatch et al. (2013) studied the approach used by Occupy Sandy in providing support after

Superstorm Sandy's impact in New York City under the Homeland Security initiative. The Department of Homeland Security (DHS) developed a comprehensive report entitled "The Resilient Social Network" under a designated task force that focuses on actions to improve community resilience to natural disasters centering on Superstorm Sandy and Occupy Sandy as an example. The main goal of the task force was to better understand emerging grassroots relief strategies such as Occupy Sandy and to make this information available to the DHS to enable the possibility of governments to work with grassroots groups in the incidence of future natural disasters. Among the many different public and private groups and organizations at different levels from the neighborhood, to the local, state, regional, and federal levels who were involved in the massive recovery efforts, Occupy Sandy "complemented these efforts and in some cases filled critical gaps" (Blachman-Biatch et al. 2013, 2). Occupy Sandy was able to successfully implement and coordinate recovery efforts in a decentralized and horizontal organization structure that provided agility and effectiveness through the development of relief hubs. Social media was a key tool and a fundamental way to attract and mobilize large groups of volunteers, help identify community needs in real time, and share information expediently. In addition, open-source software tools were used to coordinate rapid relief services. Occupy Sandy was also able to work with and deal with communities' needs, develop trust relationships, and build local capacity. The transparency in the Occupy Sandy practices with the community helped develop trust among the communities during the recovery process. The role that Occupy Sandy played also highlighted the limitations of traditional relief efforts where grassroots groups such as Occupy Sandy that were not sufficiently integrated into FEMA's "Whole Community" approach. According to the report, there was not a common operating framework to coordinate response initiatives between local, state, federal, NGOs, and unaffiliated entities during and even after Superstorm Sandy's impact (Blachman-Biatch et al. 2013). The lack of training on the use of social media was a deterrent to effective action from the government response units.

Blachman-Biatch et al.'s (2013) main goal with the report was to decrease the level of unmet needs and to improve the structures that provide relief before another disaster takes place. The recommendations also focused on improving FEMA's response coordination of activities and capabilities to prepare for, respond to, and recover from disasters with grassroots groups that operate at the local level as part of the "Whole Community" approach (FEMA 2011). FEMA's "Whole Community Approach to Emergency Management" is conceptually built as a means to collectively understand and evaluate the needs of communities by residents, emergency management practitioners, organizational and community leaders, and government officials to "determine the best ways to organize and strengthen their assets, capacities, and interests" (FEMA 2011, 3). This approach to emergency management can be proven beneficial by providing a shared understanding of community needs and capabilities taking into consideration the communities served and their cultural values. The "Whole Community" approach also includes the empowerment and integration of resources from across the community to include all members of the community into the emergency management team, the development of stronger social infrastructure, and the establishment of relationships that facilitate more effective prevention, protection, mitigation, response, and recovery activities against disasters. FEMA's "Whole Community" approach seeks to support and strengthen the institutions, assets, and networks that are already functioning in communities and that deal with community members' issues that are important to their everyday lives. The daily life structures and relationships of individuals, families, businesses, and organizations in the community are important assets to effectively act during and after a disaster. The "Whole Community" approach seeks to provide greater resiliency at the community and national levels by understanding the community's complexity, capabilities, needs, and establish relationships with community leaders, foster local action, and develop and strengthen social infrastructure, networks, and assets to their maximum advantage within prevention, protection, mitigation, response, and recovery strategies (FEMA 2011). Disaster-resilient communities are, usually, communities that function well and solve problems under normal conditions. Trust is one of the most important components for strong communities. Emergency management's challenge is to assess how to work best with different groups and organizations and to work with policies and practices based on their experience in order to improve the ability of local residents to prevent, protect against, mitigate, respond to, and recover from any type of threat or hazard, effectively. A better understanding of a community's capabilities and needs can help to efficiently use available resources despite the size of the incident or community's limitations. When there are resource and economic limitations, gathering the efforts and resources from the community can help to ameliorate budgetary constraints and pressures from government agencies, private and nonprofit sector organizations since in large-scale disasters or catastrophes, government resources and capabilities can be easily overwhelmed (FEMA 2011). The "Whole Community" approach is therefore necessary in light of the continued population shifts in vulnerable areas such as hurricane-prone coastlines. The economic development that usually follows these population shifts tends to add pressure to the coastal floodplains, barrier islands, stressing the ecosystem that supports food production, the tourism industry and suburban residential growth (FEMA 2011). The growing population of people with disabilities who live in communities rather than institutions and those who have chronic conditions such as obesity and asthma, and the communities that are confronting an increase in the senior population all add to the shifting landscape of developed coastal areas in populations whose care require different accommodations (FEMA 2011). Transportation systems and housing styles change to fit into the needs of these growing demographics and the lifestyles of the residents living in affected areas. The kinds of employments and new technologies will also have an effect in the way residents organize their leisure time and home-to-work commuting patterns. The understanding of these complexities in the communities and building on their strength are the main goals of the "Whole Community" approach to disaster recovery (FEMA 2011).

FEMA's "Whole Community" approach provides recommendations for the different kinds of partnerships with community groups such as community councils, volunteer organizations, local voluntary organizations active in disaster, community emergency response team programs, volunteer centers, State and County animal response teams, faith-based organizations, individual citizens, community leaders (representatives from specific segments of the community, including seniors, minority populations, and non-English speakers), disability services, school boards, higher education institutions, local cooperative extension system offices, surplus stores, hardware stores, big-box stores, small local retailers, manufacturers, distributors, suppliers, logistics providers, home care services, medical facilities, government agencies, embassies, local planning councils, Citizen Corps Councils, local emergency planning committees, chambers of commerce, nonprofit organizations, advocacy groups, media outlets, airports, public transportation systems, utility providers, and many others (FEMA 2011). The recommendations also included an assessment component to examine the effectiveness of the strategies implemented in the community. The recommendation from FEMA's "Whole Community" approach also included developing the capability to share awareness and understanding of the information that is available and to develop better information sharing strategies for all the groups and entities that respond to a disaster. The goal is to identify a broad base of stakeholders along with the relationships that could emerge and where the information about the community needs could be shared (FEMA 2011). The information sharing capabilities were efficiently executed by Occupy Sandy's use of social media to coordinate efforts and share information.

Despite the fact that many grassroots such as emergent response groups, voluntary organizations both secular and faith based, nonprofits, and governmental organizations response networks responded to the impact of Superstorm Sandy, the Homeland Security Report on "The Resilient Social Network" (Blachman-Biach et al. 2013) focused on Occupy Sandy and their recovery work from October 29, 2012 to August 2013. Occupy Sandy is an example of the spontaneous citizen aid that emerged after Superstorm Sandy devastating impact, from groups that had no preexisting membership, tasks, roles, or expertise previous to Sandy.

The horizontal approach to organization as an important emergent group response component was complimentary to the hierarchical centralized approach largely implemented by governmental institutions. Occupy Sandy was able to establish and deepen connections between hierarchical institutions and emergent groups. As the relevance and number of horizontal group participation grew, the unity in efforts increased (Blachman-Biach et al. 2013). Furthermore, Blachman-Biach et al. (2013) found that horizontal models of "communication, coordination, and resource management are more effective than command and control approaches" (12). It seems that even though volunteers are requested to participate in Citizen's Corp, the Homeland Security organizational apparatus seems to be challenged on how to integrate civilians to its functioning. Occupy Sandy effectively coordinated through social media the informal volunteers into a horizontal platform of operations. These volunteers were not affiliated to traditional relief organizations but through Occupy Sandy's organization they were able to offer training, conduct screening, raise funds, conduct relief activities, form tasking, and many other duties usually associated with more traditional forms of aid delivery. Within the governance of disaster, after federal responsibilities are finalized, search and rescue teams, Disaster Emergency Communication Groups, Individual Assistance Groups, Public Assistance Groups, and the other federal and state organizations complete their work on the scene of the disaster, the staff is dismissed. The federal recovery coordinators remained to continue with the provision of advice and expertise to communities, usually for years. The National Guard and the Army Corps of Engineers facilitate the delivery of services, offer technical assistance and expertise, and provide logistics planning, construction management, and management and sustainment capabilities (Blachman-Biatch et al. 2013). But it is the voluntary organizations active in disaster (VOAD) who fill in the gaps. Situational awareness is a crucial component to this effort, it entails understanding the community dynamics, identifying the stakeholders in the neighborhood as well as the needs of the specific community. The development of situational awareness is crucial and emergent response groups, largely relying on volunteers, provide support to those who are unable to prepare for a complete outage of services on a daily basis over a period of time after a disaster strikes. FEMA highlights the relevance of all groups, organizations, and individuals involved in disaster relief to know what efforts are being conducted (FEMA 2011). Community-based groups have a significant impact in helping to meet the needs of immediate survival and life-sustaining activities of the neighbors. The gaps between the needs of individuals affected and traditional government–based response along each stage of disaster recovery are usually filled by the work of emergent response groups. The services provided by the VOAD such as American Red Cross and Salvation Army in conjunction with the work of FEMA and the National Guard include services in shelter, food, medical, personal hygiene, physical reconstruction, logistic management, transportation management, child care, case management and related services, mental health and spiritual support (FEMA 2011).

Usually, efforts are duplicated because there is a lack of knowledge about the different aid provision efforts employed on the ground by different groups and lack of situational awareness especially for voluntary groups that offer aid outside of the federal or state levels. Situational awareness is less present as the response moves from the traditional state aid delivery to individuals, and neighbor to neighbor (Blachman-Biach et al. 2013). The limitations of traditional relief efforts during the recovery and relief efforts implemented after Superstorm Sandy, but likely not to be unique to Sandy, included lack of sufficient integration of emergent grassroots initiatives, like Occupy Sandy, into FEMA's "Whole Community" approach. FEMA lacked an overarching, general operating knowledge of what was occurring on the ground among local, state, federal, NGOs, and unaffiliated and extra-official groups and entities to better coordinate response efforts during and after Sandy (Blachman-Biach et al. 2013). They were not completely knowledgeable about who needed help, how to connect those in need into response efforts, updates on what was actually accomplished, knowledge about what resources and capabilities were on their way, and how to get more resources as they were needed. The Community Emergency Response Teams (CERT) with which FEMA-trained volunteers did not take into consideration how social media could have been used to facilitate the recovery process. The unanticipated impact of a storm like Sandy quickly exceeded the capacity of services allocated by FEMA Voluntary Agencies Liaison (VAL) to perform critical mission functions (Blachman-Biach et al. 2013). Voluntary organizations that were larger than the grassroots groups quickly overwhelmed FEMA VAL's time and attention, often excluding smaller grassroots groups. The urgency of the situation also prevented FEMA VALs to develop relationships with leaders and relief partners from the communities affected, that in addition to the hierarchical approach that characterizes government response generally and FEMA specifically, made the learning curve more challenging when trying to integrate services with local agencies (Blachman-Biach et al. 2013).

Under those circumstances, a collaborative system that uses social media such as the web and smartphones could facilitate coordination and enhance situational awareness without the hierarchical oversight that tends to distinguish bureaucratic governmental response, largely based on command and control (Blachman-Biach et al. 2013). In a chaotic situation, the command and control approach tends to be outmoded and rigid. Organizing a digital common platform facilitates the sharing of information about response and recovery actions in a more flexible and comprehensive manner especially in hardest hit areas with limited access to resources. Occupy Sandy was able to act expediently and effectively filling in the gap left by governmental institutions, in a flexible manner under a dramatically exhausting situation. The recovery response work that Occupy Sandy coordinated was largely composed by volunteers who were young, educated, knowledgeable about technology, had time availability, and the willingness to help (Blachman-Biach et al. 2013). The group of volunteers in the Occupy Sandy group did not have leaders per se, bureaucracy, regulations to follow, predefined mission, charter, or strategic plan. Their attitude of not saying no to requests provided an openness in their approach to best serve those who needed assistance by keeping the interactions open to ideas and suggestions (Blachman-Biach et al. 2013). In what follows, there is a more detailed description of how Occupy Sandy functioned within the challenges posed by Superstorm Sandy (Blachman-Biach et al. 2013).

According to the Homeland Security Report (Blachman-Biach et al. 2013), an operation hub and storage space was set up in Red Hook, Brooklyn, called the Red Hook Initiative where neighborhood volunteers, community organizers, individuals affiliated with Occupy Wall Street movement form a coalition with a local nonprofit, and councilwoman Christine Quinn was also involved. Volunteers worked on developing a 72-hour preparation plan after a hurricane impact with flooding and to know what to do by the hour. Occupy Sandy also set up a webpage with information on how to sign up to receive updates, volunteer opportunities, locations for donations of goods, and a list of emergency shelters. The response received quickly overwhelmed the capacity of the space with donations and the processing of volunteers. The St. Jacobi Evangelical Lutheran Church, Sunset Park, Brooklyn became the main distribution hub by October 31, 2012. In need of still more space, the St. Luke and St. Mathew Church in Clinton Hill, Brooklyn materialized through the assistance of the Occupy network. A kitchen, distribution center, and volunteer processing and training emerged at this location by November 3, 2012. Not long after, three additional hubs opened in the Rockaways and Coney Island in Brooklyn and in Staten Island (Blachman-Biach et al. 2013). Along with these efforts on the ground that coordinated the work on these sites and the volunteers work in New York City hard hit areas, volunteers with technological knowledge developed Google docs, Facebook, and Twitter accounts (@OccupySandy), managed WePay account for money donations, and updated the Occupy Sandy webpage (Blachman-Biach et al. 2013). In approximately one week after the storm 700 volunteers were working along the Occupy platform and serving food, obtaining likes on Facebook and Twitter followers in large numbers (Blachman-Biach et al. 2013). To handle the volume, 15 volunteers were managing the work from the Facebook page a week after the storm hit. Without disaster relief experience, the Occupy network was coordinating the effort of more than 60,000 volunteers. Social media outlets like the New York Magazine and The Huffington Post exhorted readers to volunteer and described Occupy Sandy as an organization that by mastering social networks provided a solid relief model for the twenty-first century (Blachman-Biach et al. 2013). The fact that there was not power in many areas meant that there was limited access to TV and Internet, no work and no gas, likely encouraged people to volunteer to the Occupy efforts announced through social media. The Occupy Sandy arrangement was flexible, volunteers chose how to help, and there was no restriction on how the volunteers might creatively fulfill a specific task based on what they thought was needed to be done by using their intelligence, strengths, skills, as well as common sense (Blachman-Biach et al. 2013).

Occupy Sandy was able to provide Just-in-Time Volunteer Training and coordinate direct aid for food and water and other resources distribution, medical care and mold remediation for those who needed it, canvassing for detecting and prioritizing needs, helping rebuild, offering legal assistance, media documentation, and psychological help, as well as community building and funding (Blachman-Biach et al. 2013). In addition to the original relief hubs, Occupy Sandy also deployed assistance to the areas they considered that might be neglected such as low-income and immigrant communities in the Rockaways, Canarsie, Sheepshead Bay, Bay Ridge, Gerritsen Beach, Long Island, and across New Jersey and a 50,000 square-foot warehouse in Coney Island (Blachman-Biach et al. 2013). Occupy Sandy also sought to help those members of populations that remain distant from official relief workers such as the elderly, the disabled, homeless, undocumented aliens, and residents in New York City Housing Authority Housing. According to Make the Road New York, 78 percent of immigrants in New York City did not request assistance from disaster relief funds and lived in disaster delineated areas (Blachman-Biach et al. 2013). Aid was delivered on the spot and as people arrived to the Occupy Sandy and other organizations community relief hubs in the Rockaways asking for help a conversation ensued to best evaluate and direct the aid.

THE OCCUPY SANDY MODEL

According to the Homeland Security Report (Blachman-Biach et al. 2013), Occupy Sandy used an organizational model mostly based on the larger Occupy Wall Street movement that rested on a horizontal organizing model that utilized consensus-based decision-making along the values of mutual aid, risk taking, and a political ideology that sought to increase awareness on a specific issue by encouraging individuals to investigate, understand, act, and change something they consider needs change[1] (Vasudevan 2015; Feuer 2012). A general tenet of the Occupy movement that was also part of the Occupy Sandy approach was to shed light on the issue of income inequality (Blachman-Biach et al. 2013). Occupy Sandy was an effort that grew from the ground up. A constant process of trial and error, understanding mistakes, meritocracy, and building consensus guided their approach. The group did not have assigned leaders per se but encouraged leadership in each person by promoting a "leaderful" attitude as volunteers who got work done. The organization of the group was supported by a horizontal, lateral approach rather than hierarchical forms of organization. Occupy Sandy's organizational approach was highly successful during the first weeks after Sandy impacted New York City because it allowed for speedy assistance. The atmosphere generated by this approach allowed for the best ideas to germinate since ideas did not need to be filtered through a bureaucratic model or authoritarian leadership style to be approved. Therefore the ideas that emerged could be implemented rapidly, evaluated and modified according to what worked in the communities and their conditions. The removal of barriers between resources and those affected produced a network that efficiently addressed the needs on the ground as these needs emerged and in close relation with community leaders and members (Blachman-Biach et al. 2013). Social capital was built along the group of individuals who helped as informal leaders who continuously participated or shared ideas considered to be good, efficient, and productive that benefited the recovery of the communities served (Blachman-Biach et al. 2013). Since traditional response organizations were overwhelmed and lacked full proof coverage for all the areas needed, the level of flexibility allowed by the Occupy Sandy approach provided the necessary freedom for individuals to use their best judgment to locate a resource and directly deliver it to a survivor. By doing so, they reached where traditional disaster relief organizations could not. Many Occupy Sandy consensus–based decision-making processes revolved around fundraising, supporting projects, and ongoing functioning of the group toward relief efforts. Pressing decisions were made quickly. Marginalized voices were given priority and efforts were made in their attempt to hear those first over those perceived as their voices being usually heard. Their approach emphasized openness to listen, change perspective, and allow the decision to go forward as an act of solidarity rather than just unanimous agreement. The Occupy Sandy Project Spokescouncil was the body that mediated the decision-making process. They managed the network's resources by allocating funds for specific recovery projects. Their meetings were open to the public and anyone could request funding for an idea. By 2013 Occupy Sandy had an operating budget of $1,377,433.57 in collected funds and less than half have been allocated to fund recovery projects (Blachman-Biach et al. 2013). The criteria by which the projects requesting funding for relief, recovery, or resilience were evaluated included spokescouncil documentation guidelines, to be aligned with the New York City General Assembly Principles of Solidarity and Occupy Sandy, and acceptance from the spokescouncil. Occupy Sandy volunteer interactions with affected community members were encouraged to be based on a humane level and on an egalitarian footing, and support offered from the premise of solidarity, "your struggle is my struggle," one of the main tenets of Occupy Sandy (Blachman-Biach et al. 2013). Occupy Sandy's notion of empowerment of individuals and communities while providing mutual aid to vulnerable populations should be done without making them feel like victims but as a valuable community members who could also help fellow neighbors affected by similar circumstances (Blachman-Biach et al. 2013). This approach also helps to establish stronger bonds with the community and relationships of trust. The social ecological and physical infrastructure of the community is strengthened in the process of community members helping others in solidarity. The tenet of risk taking in the Occupy Sandy was present when volunteers often entered badly damaged infrastructure to assess damages, needs, and cleanup (Blachman-Biach et al. 2013). In their operations, Occupy Sandy used the Internet, Facebook, Twitter, Amazon registry, websites, Flickr pictures, videos on relief efforts, Celly and Tect Loops for text messages, Sahana open source software, Google products, WePay (main fundraising software), Mappler (geographic information system with interactive web and mobile-supported maps), Occupy SMS, Constituent Relationship Management (CRM) to keep track of the Occupy Sandy volunteers, and Analog information sharing (Blachman-Biach et al. 2013). Talented and skillful volunteers who participated in the Occupy Sandy initiative were later hired by FEMA and the American Red Cross. According to the Homeland Security Report (Blachman-Biach et al. 2013), one of the disadvantages to the Occupy Sandy model was that promises were part of the fluidity of their organizational approach and the needs of the communities served, therefore accountability under such rapidly changing circumstances was challenging. There was no protocol to follow up on promises made by volunteers to those affected by the disaster and to hold the network accountable (Blachman-Biach et al. 2013).

The findings from the report state that the Occupy Sandy efforts were agile in their functionality and their use of social media was groundbreaking in comparison to relief approaches used in the past (Blachman-Biach et al. 2013). They were able to mobilize large volunteer groups, share information, and identify real-time community needs and relief projects by rapidly coordinating efforts with their expert use of social media. Occupy Sandy was also efficient and prompt in coordinating the use of the organizational infrastructure already developed by Occupy Wall Street, within days after the storm. The transparency of their relationships and practices toward relief helped develop trust among themselves and the general public (Blachman-Biach et al. 2013). Occupy Sandy is an example of the bottom-up resilient alternative of bouncing back while providing the required assistance to communities in need.

COMMUNICATION

Preparedness to climate change events must include not only redundancy in infrastructure such as transportation and land barriers in flood prone areas that were heavily affected by Hurricane Sandy, but a set of concerted efforts to improve communication and the involvement of communities. According to Miller and Rivera (2015), information flow is the first step in building community resiliency as social capital. Lin (2001) adds that information, influence, social credentials, and reinforcement as components of social capital improve adaptation outcomes. Flow of information to people and groups presents the opportunity to be exposed to different points of view and to opportunities that might not be available otherwise. Likewise, communication helps to bring the issue of climate change to the public, and attract their attention, which is crucial to unpacking the current dislocated awareness to climate change. On the one hand, scientific studies are abundant and they demonstrate the connections between human activity and changing climate. On the other hand and despite the ample evidence of climate change impacts, there is yet no full acknowledgment that translates into everyday life actions and practices for many individuals. There seems to be a disconnect between acknowledging climate change and translating it into everyday practices conducive to change. Communicating information about climate change in communities through conversations and meetings can be a critical component to improve preparedness and recovery practices.

The recommendations suggested by Homeland Security in their study of Occupy Sandy highlight the importance of FEMA to develop an efficient coordination of response actions and capacities with local grassroots groups involved in relief efforts (Blachman-Biach et al. 2013). Recommendations included developing guidelines on how to work with social media by adding it to the CERT training for preparedness and response to disasters. In addition, ongoing evaluations of how to better align efforts with the "Whole Community" approach should be embedded in the preparedness and relief efforts. Among the actions suggested by the Homeland Security study of Occupy Sandy, relief efforts should be reviewed and accounted for in order to prevent duplication of strategies by other governmental organizations. Along those lines the report also suggested operational requirements to justify and document the way(s) a specific capability is needed to facilitate the identification of gaps and to document the emergence of groups that are filling those gaps (Blachman-Biach et al. 2013). By using social media, and following the example of Occupy Sandy, FEMA should built a communications platform where mapping, community news, virtual Emergency Operation Centers (EOC), as well as relevant laws, regulations, and policies can be displayed and be easily accessible through the Internet. Technical evaluations could also provide up to date knowledge of emerging technologies that could be effective to communicate with different groups and to improve recovery efforts on the ground. The responsibility for the maintenance of the website content and technical management needs to be secured for ongoing improvement. Testing and evaluating the functionality of the system, preferably in real-life scenarios, need to also be embedded in the process of developing a social media component to recovery efforts. These recommendations could help move the emergency management closer to community needs and strengthen the resilience of communities affected by disasters in a way that more volunteers could respond in a smarter and more effective manner like the one implemented by Occupy Sandy and the relief hubs (Blachman-Biach et al. 2013).

COMMUNITY RESILIENCE

There is a need for governance at the community level with participatory structures that effectively deal with social exclusions. Use of and consideration towards indigenous knowledge, diversity in the population and wider participation in decision making processes, as well as flexibility within cultural traditions and networks are some of the aspects emphasized in the litarature that need to be taken into consideration regarding community involvement and participation for climate change adaptive practices (Tompkins and Adler 2004, Berkes and Jolly 2001).

Networks of engagement are evaluated also in terms of the support they offer to participants in vulnerable positions. Sharing information and feedback is most important to keep the channels of communication open. However, preparedness needs to include information dissemination and raising awareness about adaptive practices. Bierbroek et al. (2010) examined Europe's adaptation capability to climate change and recognized that a climate event involves "political will, human and financial resources, good coordination between key actors in different sectors and at different administrative levels, and compatibility with other policies" (15).

Similar to the efforts to engage community with efficient strategies like Occupy Sandy, FEMA's "Whole Community" approach that places relevance on social redundancies is a fundamental component of resiliency strategies. Dimmer and Lindberg (2014) studied the town of Ishinomaki in Japan, devastated by the tsunami in March 11, 2011. Dimmer and Lindberg use the example of a grassroots-led, bottom-up organization of the Ibasho Café managed by elderly community residents for victims of the tsunami (2014). The group of elders were involved in the planning, design, construction, management, and operation of the Café. The Café provided members of different communities—who ended up scattered and displaced after the tsunami—with much needed opportunities to meet, developing resiliency by fortifying social relations. According to Dimmer and Lindberg (2014), provision of places for social engagement proved to be critical for resilience.

Figure 4.3 A Concession Stand along Rockaway Boardwalk, New York City, 2017. Photo by J. Nevárez.

Figure 4.4 An Example of Public Space along the Rockaway Boardwalk, New York City, 2017. Photo by J. Nevárez.

The famed Japanese architect Toyo Ito and colleagues became involved in designing places that could offer this kind of opportunity to Japanese tsunami survivors. Ito's "Home-for-All" Project provides homeless and survivors with an informal meeting point for the community. Bonding social capital such as the one supported by Ito and Occupy Sandy enhances connectivity among community members and extra local individuals. According to Aldrich (2012), the government should be more involved in developing and supporting local community networks and social capital in addition to efforts to improve the physical infrastructure. The emphasis on social capital and community involvement with governmental actors would strengthen relationships of trust by developing a governance approach that is in tune with the communities' recovery struggles and challenges. Thus, the emphasis is then on collaboration and solidarity to build civic capacity and place governance (Dimmer and Lindberg 2014). Dimmer and Linderberg (2014) used mapping techniques to document social innovation and community resilience to highlight the critical importance of "soft" social capital for "a higher survival rate in a disaster situation and for a more positive post-disaster recovery and community development" (263). Ng and Ren (2015) also studied the use of mapping techniques to document climate change issues as a tool for urban planning. Ng and Ren included many cities around the world as case studies (Germany, Brasil, Vietnam, China, and Japan among other countries).

Dimmer and Linderberg (2014) found that community resilience facilitates public interest and civil society activities that benefit communities' disaster survival and the coordination of relief, rebuilding, and adaptation to change working along nonprofits. Tools that allow mapping the social landscape are then of crucial importance for the coordination of relief efforts, since most maps leave some communities visually underrepresented, fragmented, and poorly understood visually. Defining resilience as the flexibility to adapt to a changed and new environment and redrawing social relationships based on new social conditions, social capital is just as important as the urban infrastructure (Dimmer and Lindberg 2014). Accordingly, cities are never completed. They are a work in progress and undergo dynamic changes, with ever evolving environmental and social risks (Dimmer and Linderberg 2014). Giddens (2011) also stresses the point that coastlines will certainly change and dealing with this uncertainty requires "citizens, business, and groups in civil society to act as positive agents of environmental change, rather than simply preventing destructive acts" (121). While the emphasis on community and social resilience is extremely relevant, Miller and Rivera (2008)

Figure 4.5 Rockaways Boardwalk Benches Ocean View, New York City, 2017. Photo by J. Nevárez.

also point at the fact that fragmented communities and communities that lack trust on governmental agencies makes the process of strengthening social capital more challenging. A community is composed of a sense of common purpose among its members, rests on the assumption of mutual responsibility, acknowledgment of interconnectedness among its members, mutual respect for difference, and a commitment for the well-being of each other and to the integrity and well-being of the group (Wood and Judikis 2002; Miller and Rivera 2008). However, in most communities elements of social disruption and lack of consensus about environmental degradation can be challenging topics to champion and galvanize the required support for collective action. These are factors that affect the capacity to develop social capital. For communities where the cycle of distrust toward the government, especially when the most visible intervention by the government during a disaster is militarized, the Occupy Sandy approach is proven to be most applicable and relevant to achieve community resilience. As a critical component to communal and governmental efforts, regional and urban initiatives need to also be addressed to provide preparedness strategies to absorb extreme climate events changes at the multi-scalar levels of aid governance.

RESILIENT INFRASTRUCTURE

The overlooked importance of social capital in the form of community resiliency should not substitute for the pressing needs to strengthen and develop in an adaptive manner the physical infrastructure. These include buildings, roads, transportation, electric power, and insurance. According to Haraguchi and Kim (2015), many countries including the United States, Germany, Bulgaria, and Australia acknowledge the interconnected risks of critical systems failure. Public infrastructures and utilities are key components to the operations and business activities of a city, along with efficient urban transportation systems, an educated workforce, and reliable public services. It is key to reduce disaster risks from critical infrastructures to produce desirable outcomes for the public and private sectors to maintain productivity and the lifestyle of the urban populations. Different structural changes had to be implemented as part of the complex recovery process post Sandy. According to Haraguchi and Kim (2015), Con Edison, the electric service provider for New York City, produced a long-term plan that focuses on fortifying the electric, gas, and steam systems, diminishing the time to restore power after an outage, and improving planning before and restoration after a storm as well as enhancing the communication to customers and other relevant constituencies. According to Flemming (2014), in 2013, Con Edison spent approximately $105 million on resiliency measures and through 2016 plans to spend $1 billion (Consolidated Con Edison of New York Co 2013). Con Edison plans to redesign underground networks, flood-proof vulnerable facilities, and invest in more smart-grid technologies (Consolidated Con Edison of New York Co 2013). In order to facilitate rebuilding and replacement of damaged transportation equipment and facilities, the US Department of Transportation awarded approximately $886 million to the New York Metropolitan Transportation Authority (MTA) after Sandy (US Department of Transportation 2014). About 60 percent of $886 million ($535 million) is assigned for repairs that are critical to three under-river subway tunnels—Greenpoint, Montague, and Steinway. The second largest portion of about $138.9 million (16%) is assigned for the repair of the Long Island Railroad and Metro-North Railroad substations and power infrastructure (Haraguchi and Kim 2015).

Regarding adaptation, insurance innovation is a key issue, especially when the state's role as an insurer will be under great strain. Insurance claims are widespread mostly in developed countries since that industry is most developed there. This is a key issue especially in areas where the insurance structure barely exists and in some cases does not account for the kind of recent extreme weather impacts that after Sandy are exemplary in areas like Houston and Florida with Harvey and the Caribbean with Irma and Maria. In the context of the United States, however, insurance does not cover, for instance, damage to roads, railways, and other public infrastructure. Hurricane Katrina's insurance losses amounted to almost $49 billion much less when compared to estimated total losses of $144 billion. Insurance losses for Hurricane Sandy were $35 billion (Freedman 2013). According to Giddens (2011), weather-related risks create enormous problems both for private insurers and for the state. A measure of successful adaptation will depend on how the private insurers and the state can develop appropriate insurance policies.

ALTERNATIVES IN INFRASTRUCTURE: DESIGN AND ECO-DESIGN

Rebuilding after a climate event like Sandy requires an analysis of cost-benefits, base flood elevations, and FEMA compliancy issues largely based on damage to property and public space considerations. Approaches to flooding are diverse, while most focus on deterrence by building dams, levees, and walls, others focus on a more ecological-based design that uses natural landscapes to absorb and deter the impact of floods on property and public space. Other cities are well known for their protection against storm surge and have already taken steps to prepare for changing weather patterns that include flooding. The London Thames Barrier has the world's largest movable flood barriers.[2] In the Netherlands, the Delta Works Storm Surge Barrier, including the Eastern Scheldt, is a complex group of storm surge barriers designed to deter flooding.[3]

In New York City, a group of resident architects carefully examined the challenging factors that came into play after Sandy, not only in terms of housing redesign and coastal environments recuperation but drawing from their personal experiences with Hurricane Sandy. They developed the "Comprehensive Coastal Communities Playbook." Horn (2013) analyzed and proposed different housing and public space designs for flood prone areas to address the impact of Hurricane Sandy in their communities. Design competitions also surfaced and the Rebuild by Design Hurricane Sandy Competition in 2013 was sponsored by the US Department of Housing and Urban Development in partnership with nonprofits and the philanthropist sector, attracting many firms to participate.[4] The designs included development of the infrastructure to provide areas for public space that could easily adapt to storm surge and flooding while maintaining diverse uses when there was no flooding. Some of the designs also addressed intervention to the ecosystem per se by proposing wetlands, sand dunes, and other natural barriers to flooding. The Rebuild by Design initiative is now partnered with 100 Resilient Cities and seeks to provide resiliency projects that are participatory with the communities.[5]

For the Superfund site in Brooklyn Gowanus Canal,[6] Eric Olsen designed a set of platforms that tolerate water surge and transforms its uses allowing for the creative implementation of plant beds that are resistant and become robust with water.[7] The design of housing for flood prone areas like those developed for Australia and Bangladesh include sustainable and cost efficient alternatives. The Dion Seminara Architecture Firm won the Award for the L. J. Hooker Grantham Affordable Flood Home Design Competition, 2011, for Australia's flood prone areas.[8] The house is easy to clean up after the flood and provides flexible use of the elevated ground level for storage and cars. It is light, easy to transport, and expandable by interlinking pavilions, and includes platforms for solar panels and hot water systems. The materials used are sustainable and environmentally friendly. The Lift House designed by Prithula Prosun uses an eco-friendly low-income flood-proof technology that provides sustainable housing for low-income communities in Dhaka, Bangladesh.[9] Rainwater is collected, filtered using a service spine, and recycled through bio-sand filtration. Nutrients for the garden are dropped via an underground pipe system that collects urine. The shared composting toilet can create compost from human waste to be sold or applied to the vegetable garden after 10 years of use. These alternative designs for houses are largely the result of rebuilding with climate change in mind without losing the opportunity to make obvious connections evident. By openly addressing issues of climate change such as Australia's government department of climate change, experts can study and produce reports from which policies can be enforced. This will provide a much needed shift in focus and approach from the way infrastructure has previously been conceived. Outdated infrastructure could be replaced with more organic, adaptable, and intelligent ones. This initiative to inform and link climate change to infrastructure could be supported by higher education institutions in New Jersey and New York. Considering alternative housing options that are adaptable to changing weather conditions and developing communities of support to face the social and physical challenges of climate change seem to be the best path for the reconstruction and rebuilding of resilient landscapes after Sandy. The Saltmeadow in Far Rockaway designed by JDS Development Group is also an example of housing design that takes into consideration water surge.[10] This design however is on the upper end of luxury housing.

Norman Foster designed a prototype for Red Hook Point of the Old Revere Sugar Factory, expected to be completed by 2020, where the structure is well above projected surge levels.[11] SCAPE designed Living Breakwaters, a coastal green infrastructure project with the purpose to reduce erosion and damage from storm surge.[12] Living Breakwaters is a more ecological approach to intervene in the coastal ecosystem. Likewise, the architecture firm WXY architecture + design collaborated with local and federal agencies to build the elevated 5.5-mile concrete boardwalk in the Rockaways from Beach 9th to Beach 126th street in a sand-colored concrete precast planks atop newly planted sand dunes that offer protection from a possible flood tide while also provides the traditional passive leisure characteristic of boardwalks along the eastern seashore. The ramps from the street are sprinkled with recycled glass and the ramps to the beach are built from recycled plastic lumber to maintain the stability of the dunes underneath.[13] After five years, the redesign of the Rockaways Boardwalk has proven to be an excellent combination of forward-thinking regarding possible storm surge in the future due to climate change and the provision of a pleasant beach environment with multifunctional structures along the boardwalk for bazaars, music, and food options. More designs have been developed to address urban farming soil pollution[14] academic Bloomberg Center, Cornell Tech Campus, Roosevelt Island,[15] and retail centers such as the Empire Outlets Mall in Staten Island[16] and power hubs such as Resilient Power Hubs.[17]

Both social and physical redundancies are necessary steps to recover from extreme and slow-onset climate impacts. Policies that support preparedness, mitigation, and adaptation will be discussed in the following chapter.

NOTES

1. http://occupysandy.net/.
2. http://www.environmentandsociety.org/arcadia/thames-barrier-londons-moveable-flood-defense.
3. http://www.deltawerken.com/the-oosterschelde-storm-surge-barrier/324.html.
4. http://www.rebuildbydesign.org/.
5. http://www.100resilientcities.org/.
6. https://www.youtube.com/watch?v=gzYv3NKxH4U.
7. http://www.orliplus.org/people.html.
8. http://www.worldarchitecturenews.com/project/2011/16882/dion-seminara-architecture/lj-hooker-flood-design-competition.html.
9. http://www.worldarchitecturenews.com/project/2011/16882/dion-seminara-architecture/lj-hooker-flood-design-competition.html.
10. http://jdsdevelopment.com/saltmeadow/.
11. http://www.scapestudio.com/projects/red-hook-point/.
12. https://stormrecovery.ny.gov/living-breakwaters-tottenville.
13. https://www.dnainfo.com/new-york/20170526/rockaway-beach/rockaway-boardwalk-complete-hurricane-sandy-rebuild.
14. http://www.added-value.org/whatsnew/.
15. https://www.designboom.com/tag/morphosis-architects-thom-mayne/.
16. http://www.empireoutletsnyc.com/news2016.html.
17. https://www.nycedc.com/blog-entry/rising-challenge-bright-powers-resilient-power-hub.

Chapter 5

Climate Change

Resilient and Redundant Social Policy

The kinds of risks and vulnerabilities posed by climate change affect different groups in society differently. The necessary strategies for resiliency and adaptation to mitigate and recover from extreme climate change events as well as slow-onslaught require a trial and error, experimental approach that is situated at the level of community actors' and stakeholders' awareness from the specific locales affected, as well as their understanding of the governmental administration of aid. The degree of uncertainty and vulnerability that communities are exposed to by climate change varies and so should the mitigation and recovery strategies developed to address preparedness and adaptation to climate change impacts.

As a complex issue, a comprehensive approach to climate change preparedness, mitigation, and recovery is critical to address all the interconnected components to the management and administration of aid. Under neoliberalism, cities are pivotal stages for the development and implementation of resiliency policies. Cities are increasingly becoming autonomous and powerful actors in the advocacy and implementation of climate change prevention, mitigation, and response measures. Neoliberal urban governmentality functions at a multi-scalar level, which includes individuals, communities, and government to absorb the social, economic, physical, and ecological impact of natural events in the context of austerity and market deregulatory practices. The governance of aid under a neoliberal landscape of government retrenchment and austerity highlights the importance of community. Climate change mitigation, preparedness, and recovery begins and ends with community, especially during the challenging neoliberal policies that curtail government spending under the austerity measures implemented. The administration and management of populations through the governmental structure in the governance of aid to impacted areas and individuals requires community involvement

Figure 5.1 Jacob Riis Beach Clock, Rockaways, New York City, 2017. Photo by J. Nevárez.

and bottom-up approaches. It also requires a care of the self that is informed by alternative ways to understand people's relationship with nature and the environment in this anthropogenic era. The constitution of a climate change subjectivity of awareness and the collaborative approach necessary to address climate impacts need to rest on a different educational system aided by social media and communication-mediated technologies where the information regarding climate change mitigation is redundant and repeated traveling through different channels and media to reach different groups in the population. In New Orleans, Miller and Rivera (2008) learned that policy based on models of citizen participation are key for the development of civic trust, sustainable development, and social equality after their experience with Hurricane Katrina, one of the strongest and intense extreme climate event experiences that showed lack of preparedness and slow recovery.

Social policy regarding climate change would focus on specific strategies that in addition to address the obvious infrastructure issues that are right away palpable after an extreme climate event would need to strengthen the complex social and community relations as a main focus for preparedness and recovery efforts. The conditions of a receptive and functional community dynamic can allow for the input of external expertise and technology transfer to motivate municipal climate engagement (Krause 2013). Many argue that civic engagement and awareness about climate change issues increase after an extreme climate event. Therefore, community functionality and cohesiveness is crucial and a necessary precondition to prompt recovery and intelligent preparedness.

According to Zimmermann et al. (2015) external expertise and information sharing are keys in the implementation of climate change mitigation innovation strategies, even under the highly experimental manner in which these strategies are developed and applied in cities globally. External could complement administrative expertise at the local level of policy development and application. This expertise, however, needs to be interdisciplinary as the character of climate change preparedness and recovery involves physical infrastructures, local, and trans-local communities in the by now generalized urban context of cities, where most of the world population resides. Policy entrepreneurs and policy transfer among cities and mayors who have largely been the leaders in the development and implementation of climate change mitigation and recovery plans require an active community rich in social capital that involves professionals from varied disciplines, stakeholders, and residents. Kemmerzell (2016) also argues that innovative local climate protection policies are necessary. Kemmerzell advocates for a tripartite approach that includes the invention of policies, after the fact evaluation of innovative policies implementation impact, and the diffusion of innovations information and knowledge as a component of policy transfer (2016). Process-oriented case studies can provide a systematic approach to the description and analysis of policy creation and implementation. Mapping after an extreme climate event has shown to be extremely valuable to account for the impact on property as a preventative measure to assess climate security vulnerability by focusing on situations where large sectors of the population could be at risk of death from climate-related hazards (Werrell and Femia 2017). The factors that Werrell and Femia (2017) identify in their model include four "baskets" to indicate the "physical exposure to climate hazards, population density, household and community resilience, and governance" (124). For the recent Hurricane Maria that devastated Puerto Rico, a group from the Center for Spatial Research at Columbia University were providing assistance via crowdsourcing by developing maps of the affected areas in the island to include elements of the infrastructure that were affected such as buildings and bridges.

The diffusion of climate mitigation policies involves policy transfer among different localities, and rests upon learning from best practices and their adaptation to local circumstances (Kemmerzell 2016). The role of political institutions, civic engagement, public participation, as well as building and mobilizing capacity in practice are key components to the implementation of climate-related policies that compete for resources, leadership, and public attention (Hughes and Sarzynski 2015).

In this regard, transnational and trans-local networks of information sharing need to be generated, as they are, in the form of websites, conferences, and social media and town hall meetings or other kinds of meetings where the local community can be involved directly. The approach suggested can benefit the identification of best practices in different physical, social, and economic contexts where situational awareness and whole community approaches can be matched to the policy transfer accurately and to the particular needs of community constituencies. Hawkins et al. (2016) mentioned three different explanations for commitment that even though directed toward sustainability, they shed light on the different local and community levels issues that climate change preparedness and adaptation policies also need to involve. These explanations of local commitment are "local sustainability priorities, regional governance, climate protection networks, interest group support, local fiscal capacity, and characteristics of the local governing institution" (Hawkins et al. 1902, 2016). They found that local priorities, participation in regional governance, and membership in climate protection networks influence the likelihoods of cities' commitment to sustainability. According to Hawkins et al, local interests refer to business and chambers of commerce interested in growth and land development, groups that are interested in environmental protection and natural resources conservation, as well as organizations at the neighborhood level that are inclined to support and promote policies that protect perceived quality of life to maintain or increase home values (2016). The development of websites that contain copies of municipal climate protection indexes as well as mitigation, adaptation, and recovery plans created by different municipalities and cities, such as those developed by New York City and the Gulf Coast, help construct a policy transfer archive that is cross country and international. These archives can include best practices and innovation for local climate protection (Kemmerzell 2015; Krause 2011). Academics, researchers, and practitioners should be an important component of this policy transfer effort in addition to local, grassroots representatives. Initiatives that emerge from concerned professional individuals are also necessary—especially those that combine a different set of skills from professionals of different disciplines concerned with the urban environment and climate change. These professionals include architects, urban planners, landscape architects, sociologists, economists, artists, and professionals from other related disciplines. In building capacity and capability for protection, mitigation, response, and recovery, the National Response Framework of Homeland Security (2013) seeks to implement policies that are scalable, flexible, and adaptable. Their principles include engaged partnership with stakeholders, tiered response, operational capabilities that are scalable to fit diverse sets of scenarios. They also include readiness to act in confronting risk hazards such as hurricanes, earthquakes, wildfires, and floods among other risks.

According to Giddens (2011), climate change policy should be anchored on the participation of the state to enable different groups in finding solutions to problems that affect many. Especially, according to Giddens, these groups should be those operating from the bottom up (2011). The role of the state and/or government should focus on making sure that there is visibility to accepted local implementations of climate change mitigation, adaptation, and recovery that are effective and successful for the community as a whole. Giddens also suggests that there should be a convergence of the different areas that are affected by extreme climate events and climate change more generally, in the short and long term along a wide spectrum of areas of concern that affect urban living (2011). For instance, these areas can include departments concerned with urban planning, energy security, community outreach, technological innovation, lifestyle, and urban politics (Giddens 2011). Progressive policies on climate change mitigation, preparedness, and recovery that incorporate low-carbon technologies, business practices, and lifestyles should partner with governments, businesses, climate change mitigation advocates, scientists, and researchers in a collaborative attempt to restructure "the capitalist political economy along more environmentally (climate change mitigation) lines" (Giddens 2011, 73). Giddens' reformist approach is debatable since it places much trust in the government as a leader in recovery efforts, certainly the role of the government and its involvement in climate change mitigation practices at different scales, is key. However and even though the state role and more specifically governmental institutions at the city level are crucial in the development and implementation of climate change mitigation and adaptation policies, without customizing to specific communities and ensuring community involvement, governmental leadership could prove inefficient if it remains highly bureaucratic under conditions of austerity measures and precarity. Foregrounding climate change positives can keep global warming and climate change at the forefront of the political agenda. Climate change positives refer to those aspects already accomplished by different local efforts that can help guide awareness about climate change recovery and mitigation strategies. This can be useful communication content to galvanize support and action. For instance, after 10 years of collective efforts by professionals and volunteers to repair sand dunes in the Atlantic coast of Puerto Rico and Florida, in 2015, a record-breaking hatchling of turtles took place. Currently that same area is being restored after the devastating impact of Hurricane Maria on September 20, 2017. Pictures of the New

York City skyline before and after carbon emissions showed a stark difference and improvement in the quality of the air due to climate change mitigation policies implemented by the city. These kinds of informational pieces can be used as a great tool to develop support for climate change mitigation policies but also to develop awareness and strengthen solidarity within communities themselves and their biota. Likewise, the percentage principle helps illustrate that any course of action taken regarding policies involves a certain element of risk and opportunities, and those should be considered to their full extent (Giddens 2011). Overdevelopment in many parts of the world is part of the entrepreneurial city global trend, building construction for a growing population functions as a way of maximizing the multiplier effect of cultural institutions, lifestyle options, private businesses, and residential development usually targeting high-income dwellers. This kind of development contributes to gentrification and what Harvey refers as "accumulation by dispossession" (2004). Alternatives to overdevelopment in low-lying coastal areas is a significant step in diminishing the worst impact to vulnerable areas. Pre-emptive proactive adaptation to climate change mitigation policies based on the assessment of risks that include the constantly evolving scientific information available is a relevant component of technology and knowledge transfer.

The 100-Year Flood Map in New York City that among its many functions provides evacuation routes and flood risk, and is also used to determine insurance premiums. However, this tool is outdated because it wrongly assumes extreme climate events to happen sporadically (every 100 years) when extreme climate events are showing to be more continuous and intense and some argue that they might become as frequent as every five years after 2030. Catastrophe bonds by insurance companies could cover the costs of the poor in developing countries working in conjunction with non-governmental organizations (NGOs) and foundations to provide necessary insurance coverage in areas that currently face recovery without a strong insurance platform (Giddens 2011). Extending the role of insurance companies to the developing world is of utmost priority to address the challenges of climate change impacts. The problems, policies, and politics of climate change mitigation, preparedness, and recovery require a more horizontal organizational approach to the sharing of information and technology transfer across actors and scales of governance. NGOs can also act as pressure groups to bring climate-related issues to the fore and play a significant role in coordinating scientific information to share with decision-makers and the public at large, most importantly. The United Nations also plays a pivotal role in coordinating efforts and sharing information regarding climate change impacts mitigation, preparedness, and recovery. The number of agreements and offices that address climate change issues have increased under the United Nations organizational umbrella and leadership. Subsidies for industries promoting energy-efficient practices need to continue increasing. Likewise, carbon taxes to polluting industries will shift the burden of climate change from individuals. The Natural Resources Defense Council is gathering enough scientific information to allow for private individuals to suit polluting companies. Direct regulation like those observed in major cities like New York City has brought buildings to code, energy efficiency standards, controlled land use, and the labeling of products to account for those that are eco-friendly. The calculation and report of emissions by different countries globally seek to provide, as stipulated in and recorded by many agreements, a level of accountability and transparency to the efforts to mitigate climate change to which more countries are increasingly participating. Even in countries that refuse to participate in climate agreements, cities are stepping up to reify those accords in their local capacity and continue to support the development of climate change policies.

The policy changes necessary to mitigate climate change impacts should not be based on energy security only but climate change mitigation per se. Individual policies are usually not perceived unless they are combined with other initiatives simultaneously, their impact tends to be aggregated and related. The governance of aid as well as of the implementation of climate change policies makes of the government a coordinating body that manages initiatives within the context of the different risks, political landscape, economic priorities, and interventions to galvanize support for climate change policy. The "polluter pays" principle could be an effective tool that, if sufficient support could be developed, can provide a balance for the zones that are unevenly affected by climate change impacts (Giddens 2011). The economic and fiscal framework to move toward low-carbon emissions is also part of the government's responsibility at any level. The multi-scalar character of the governance efforts required for the administration of aid and the development of climate change policies requires the inclusion and integration of local, regional, national, and international components. Strong national programs, international agreements, and collaborations are necessary in order to address climate change preparation in a comprehensive manner (Giddens 2011). These agreements have not yet reached the best outcomes but engagement and acknowledgement about their importance is evident.

In this regard, the NYC Mayor's Office of Recovery and Resiliency Interactive Map[1] identifies the rebuilding of structures affected by Hurricane Sandy. The opportunity to develop resilient community centers where information about preparedness and mitigation can be obtained along with a virtual platform to help coordinate efforts should not be missed. Public space platforms where communities can organize efforts to develop resilient cultures and communities of solidarity can benefit from the relief hubs model developed and implemented by Occupy Sandy. The public forums can be a valuable inclusion to the Recovery and Resiliency Interactive Map (which currently showcases mostly infrastructures) to provide a platform for engagement with the public along climate change concerns. The climate change community centers can also offer opportunities to resocialize communities to live with nature in relationships of cooperation and collaboration based on an urban education with ecology and nature at its core. Scenario planning that models possible impacts and plans for preparedness and recovery can be a useful tool to resocialize our sensibilities toward a relationship with nature different to the modern, industrial, capitalist notion of nature as a resource to be exploited (Werrell and Fermia 2017). Another valuable tool regarding climate security discussed by Werrell and Fermia is "dark reports." Accordingly, these reports can include what is not known about preparedness and to communicate this information to communities and planners alike to help anticipate potential risks associated with the uncertainty and vulnerability to extreme climate events (2017). Examples of these "dark reports" could be a mapping of polluted areas, weak infrastructure, hospitals, schools, sewage treatment plants, and other relevant information for communities with high risk. Zomer and Hsu (2015) also recommend land use planning and transportation design as good opportunities to integrate climate planning to the city, regional, and national scales. The role of the community can also include cadres of professionals who are involved in developing climate change awareness and alternatives. After Hurricane Sandy a group of local architects who lived in affected areas developed the "Comprehensive Coastal Communities Playbook" by the Operation Resilient Long Island to address alternatives in building and public space design. Currently, in New York City, some of the most resilient structures are museums such as the Whitney Museum that even though located in a flood prone area, its design took into consideration flooding and sea level rise. Efforts should also be made to include a level of accountability for air traffic and shipping. Both are considered big polluters but they remain unaccountable due to the international character of the areas through which they move. Already celebrities are using their public persona to influence and raise awareness toward climate change.[2] As the governance of climate change evolves, a platform for climate change–related jobs in the United States and internationally is available and includes job opportunities as energy data officer, director, community management, project officer, program manager, team assistance, climate scientist, communications internship, senior project manager, business development and sourcing manager, international chief technical adviser, project administrator, policy manager: low-carbon industries, sourcing manager, green electricity, consultant energy for sustainable development in Caribbean buildings, climate finance office, expert in agriculture sector modeling, expert in transport sector modeling, senior program office, International Programs, and Institute for Sustainable Communities, among others.[3]

As the need for bold action regarding climate change accentuates, the complexity of social policy is also taking shape in a similar entrepreneurial, experimental, and spotty way as the current predominant neoliberal form of governance to absorb and adapt for change. The communities of solidarity needed to galvanize support and preparation for extreme climate impacts and slow-onslaught can be drawn from state-supported policies and can transcend climate change concerns to also address a myriad of social issues. Communities of solidarity are needed in substitution to the praise for an unfettered individualism that is generating many weak social subjects. Research along the lines of bridging the gap between climate change and everyday practices is necessary as well as a systematic examination of the notion of community that strengthens individual participation in collective action and support, rather than policing practices of social control. The notion of community that emerges from the neoliberal exaltation of the individual and our participation in communication-mediated technologies are at the core of the issues that we need to examine more fully since they carry consequences for our social engagement in communities of solidarity. These two areas in my view deserve serious attention in terms of research.

NOTES

1. https://maps.nyc.gov/resiliency/.
2. personalities.org.
3. https://community.iisd.org/jobs/.

Conclusion

Tackling Climate Change: Epistemological Rift in Governing Disaster

The Anthropocene has made evident the epistemological rift generated by climate change. Assumed notions about the relationship between nature and society dated from industrialization and modernity's ethos of progress and development produce urban landscapes of disaster. The impending epistemological change announced by the Anthropocene requires an understanding of the relationship between human, nonhuman, objects, and organic and inorganic matter different from the outdated and utilitarian one brought about by industrialization and capitalism. Instead, relationships of solidarity rather than competition and exploitation seem necessary to rearticulate and emphasize alternative conceptualizations and practices that focus on understanding climate change processes in urban areas. The ontology of resilience by which neoliberalism and climate change impacts the multi-scalar spheres of everyday life has a positive patina of experimentation where knowledge is produced by a trial and error learning process, unavoidably and necessarily. The fluid and relational conceptualization that the Anthropocene pushes forward requires an ethics of kinship and solidarity as a way of bouncing back from uncertainty, vulnerabilities, and risk with flexible adaptive strategies. The challenge rests on developing strategies that account for the global character of climate impacts that transcend national boundaries vis-à-vis climate change policies that are circumscribed to the delimited geopolitical area of the nation and the localities affected. Werrell and Fermia point out that the "cumulative effect of climate change contribution to the erosion of state legitimacy" might be becoming a significant issue and not recognized as such yet (2017, 16). There is the need for new governance structures where niche institutions, national governments, and regional and international security institutions take into consideration the consequences of climate change in planning, strategies, and operations on an equal footing. For Werrell and

Figure C.1 The Rockaways Sand Dunes Recovery Project, New York City, 2017. Photo by J. Nevárez.

Fermia climate change represents a threat to global security, especially in areas where there are unstable political circumstances as well as water and food insecurity (2017). After an extreme climate event water and food tend to become scarce and a source of stress and instability among the many other everyday life aspects that become uprooted. A challenge to security in general is the abyss or disconnect that seems to exist between the think tanks at the higher echelons of security organizations such as Homeland Security and FEMA that develop research-based policies and the kind of training troops receive to face their responsibilities on the ground which are usually limited to preserve order and protect property rather than cocreatively implement strategies to safeguard and develop community, broadly understood.

Braun (2014) emphasizes how the role of infrastructure tends to be the key focus and priority for preparedness and adaptation policies. Protecting critical infrastructure is becoming what security means when planning for preparedness with climate change impacts in mind. Thus preserving technological systems and infrastructures supports populations' productivity. Infrastructure is also increasingly an indirect form of the administration and management of urban populations as biopolitics. The overemphasis on infrastructure clouds possibilities to develop strategies that are based on alternative understandings about the relationship between humans and biota contained in an urban ecology that is increasingly upset and pushed to the limits. The overemphasis on infrastructure also undermines the importance of community participation in devising the best practices for preparedness, mitigation, and recovery from climate change events, from the ground up as a process of cocreation. Climate change social policy in a landscape increasingly portrayed as fluid and flexible carry overtones that are equally reflected in the necessary capacities of subjects within the entrepreneurial ethos of neoliberal policies of advanced capital accumulation. Working with nature and absorbing its changes characterize the kind of accommodations that are set forth to develop resilient urban landscapes. Likewise, adaptation and the ability to absorb change are crucial in shifting neoliberal policies that are based on austerity that produce precarity. Administering aid under these austerity conditions jeopardizes the response to communities with already limited resources and that might already exist at the margins of urban development. The city is therefore the bastion for a myriad of multi-scalar creative policies regarding climate change mitigation and preparedness interventions.

Nature is considered a soft infrastructure and preparedness is based on an adaptability to absorb rather than fortify against (Braun 2014). Sustainability and resiliency are becoming the current content on the discourse of global urbanism to build and discipline communities. The paradox of the city rests on the notion that it is at this scale that the city has become a problem as a product and epitome of industrialization but it also carries its own solution regarding climate change mitigation, preparedness and adaptation policy development and implementation practices (Davis 2010). Resilient urbanism is then an extension of the particular neoliberal mode of governance, a necessary acknowledgment to understand the context in which practices and strategies for preparedness and adaptation to climate change are modeled. To prepare for and anticipate future risks are constitutive of the current ideological infrastructure of neoliberal multi-scalar resiliency. Governing disaster scales down from the administration of populations to the constitution of a subject in an entrepreneurial mode of experimental interplay of different actions and practices. Producing something new is then part of maximizing existing practices to develop alternatives such as those illustrated in the Occupy Sandy model and the relief hubs that from the bottom up effectively galvanized momentum. Occupy Sandy became an alternative model that emphasized solidarity and skilful adaptation with the aid of communication-mediated technologies and a horizontal framework of command that was participatory and respectful especially of those voices that are not usually heard. In the Occupy Sandy relief hubs communities of alliances were built among different organizations, groups, volunteers, and concerned citizens as a response to extreme climate change impacts based on an ethics of kinship and solidarity. The alternative, heterotopic other spaces of gender, race, refugee, and species inclusivity can be generated within the collective and flexible dynamism of alertness and action to resignify and cocreate our current approach to circumstances: climatic, economic, social, and spiritual.

The precarity of neoliberal austerity policies impacts extreme weather-related efforts. Governing disaster includes governmental federal organizations like Homeland Security and FEMA budgets are determined by the Congress for the allocation of recovery funds largely dependent on the prevalent political atmosphere and its priorities. More recently after the impact of Hurricanes Harvey, Irma, and Maria in Houston, Florida, Puerto Rico, and the Virgin Islands, respectively, the states and territories affected would have to access funds from the same recovery budget, which has been already dwindling due to budget cuts, austerity measures, and the high quantity of consecutive climate impacts. Katrina remains the most costly extreme weather impact costing 156.38 billion in 2005, and Hurricane Sandy cost 68 billion in 2012 (Werrell and Femia 2017). The most recent hurricanes—Harvey, Irma, and Maria—in 2017, will likely surpass these figures and 2017 will remain one of the most costly hurricane seasons on record.

Regarding Homeland Security and the role of the military in directing and implementing response and recovery policies, it seems that the upper echelons of the military might have a sophisticated understanding of climate security issues, migration triggered by climate change, and strategies informed by think tanks based on research. But it is likely that the boots on the ground that are deployed to assist after an extreme climate impact might not be versed in these cutting edge and advanced strategies and knowledge. This can be an excellent area for future research—how to develop training sessions and socialize the military into different relations with nature based on collaboration to trickle the best practices with situational awareness and creative participation from the field. Likewise, the challenge remains on how to convey in everyday scenarios the connections between living and nonliving organisms as we experience the extreme climate of the Anthropocene to develop a different kind of sensibility toward living and nonliving matter including objects.

The transboundary character of climate impacts also requires a transboundary approach to policy that involves different states, organizations, and stakeholders—globally. This requires a new world order built on an international system of cooperating sovereign states (Werrell and Femia 2017). The current security landscape has to assess the ability or inability of governments to effectively manage rapid change, ensure security and prosperity for their publics, and maintain legitimacy. The United Nations has stepped up to this challenge and researchers also participate in the best practices in knowledge and technology transfer. The implementation of these practices on the ground requires taking into consideration the different stages of development of the communities affected but also an approach that is based on the situational awareness of these communities to rather work with them in anticipating risks and building relationships of trust. Constanza (2017) suggests the notion of the global commons or a global civil society where public trusts safeguard access to clean air and water in urban, forest, and maritime landscapes to which public attention and understanding are key issues.

The political ecology of landscapes left after extreme climate is a critical perspective from which to understand the interplay of social, political, cultural, and ecological factors that impinge on preparedness and recovery from extreme climate impacts. The critical natural and built capital of ecosystems including the atmosphere, oceans, biodiversity, and cities are part of the assets for a communal public good. Community-focused efforts provide the opportunity to generate social and cultural capital for recovery and long-term policy that is creative, when the community is engaged (Constanza 2017).

Regarding Hurricane Sandy, the management of aid showed lack of preparedness and it functioned as a catalyst to develop and implement many initiatives to climate preparedness and mitigation policies largely led by the city. In the Rockaways, an area badly damaged by the storm surge, sand dunes have been retrofitted and a new boardwalk as well as new buildings have been completed. The Rockaways has become fully functional after five long years. However, Murray states in his report (NJ Sandy Panel: Still Dissatisfied with Recovery: Despite Improvement, Many Survivors Still Face Hurdles Five Years Later) that from 432 participants, 55 percent are still dissatisfied with the New Jersey Recovery efforts and 56 percent say that people "like them" have been forgotten. Thirty percent of the sample were participants between the ages of 60 and 69 and 23 percent had an income under $50k. In their report "The Long Road: Understanding Sandy Recovery Home and Lessons for Future Storms Five Years Later," Devecka-Rinear et al. (2017) found that from more than 500 New Jersey participants, 98 percent of the respondents who participated in the study reported Sandy damage caused by flood waters, 40 percent do not have the funds to complete rebuilding their homes, 56 percent have had trouble paying the bills and/or affording food and gas since the storm, 70 percent reported that they had developed new physical or mental health problems or deteriorating pre-existing health conditions since Sandy, and 57 percent believe they are not better prepared for future disasters than they were before Sandy. Their recommendations include affording the ability to pay an income-based repayment plan, better policies to handle contractor's fraud, improvement in the accuracy of flood maps, insurance affordability for families facing economic obstacles, making assistance programs consistent through the recovery process, and making mitigation and disaster prevention efforts a priority (Devecka-Rinear et al. 2017). The participants also mentioned that communities need long-term health assistance and programming particularly for mental health support (Devecka-Rinear et al. 2017). An extreme climate event uproots residents and the recovery takes a long time, if ever completed.

In the Rockways, Dubois (2016) focused on restoration practices in the beach and concluded that the restoration practices do not secure public access; to the contrary, the emerging trend is the privatization of the beach intended to fit the economic real estate and other developments interests for profit, from investments in the area. Accelerated privatization is a common trend in post-disaster geographies where the residents have been displaced and private investors capitalize on the landscape restoration for profit. The communities of solidarity that emerged immediately after Hurricane Sandy's impact are necessary ideologies for a post-disaster culture that requires continuity. The notions of political ecologies and a global commons are basic components to understand resiliency efforts in general, for communities that lack resources to effectively bounce back, specifically. Climate is global, relational, and multi-scalar. Its impacts affects us all. At the juncture of the city, international agreements, bottom up approaches based on communities of solidarity and a global governance structure conducive to protect the global commons including humans and nonhumans, lay efforts to mitigate, prepare, and adapt.

Bibliography

Abid, Muhammad, Janpeter Schilling, Jurgen Scheffran and Farhad Zulfigar. 2016. "Climate Change Vulnerability, Adaptation and Risk Perceptions at Farm Level in Punjab, Pakistan." *Science of the Total Environment* 547, no. 16 (March): 447–460. doi:10.1016/j.scitotenv.2015.11.125.

Acosta, Joie D., Anita Chandra, Vivian L. Towe, Yandong Zhao, and Yangxu Lu. 2016. *Enable Non-Governmental Agencies to Get Engaged in Disasters Toolkit.* Santa Monica, CA: RAND Corporation. Accessed January 22, 2017. https://www.rand.org/pubs/tools/TL202.readonline.html.

Adaptation Fund Board Secretariat, The. 2016. "An Analysis of Climate Change Adaptation Reasoning in Project and Programme Proposals Approved by the Board." Bonn, Germany 2015. Accessed July 19, 2016. http://www.adaptation-fund.org/wp-content/uploads/2015/09/AFB.PPRC_.17.5-Analysis-of-climate-change-adaptation-reasoning_final.pdf.

Adeney Thomas, Julia. 2014. "History and Biology in the Anthropocene: Problems of Scale, Problems of Values," *American Historical Review* 119, no. 5 (December): 1587–1607. doi:10.1093/ahr/119.5.1587.

Adger, Neil W. 2000. "Social and Ecological Resilience: Are They Related?" *Progress in Human Geography* 24, no. 3 (September): 347–364. doi:10.1191/030913200701540465.

Adger, Neil W., Nigel W. Arnell, and Emma L. Tompkins. 2005. "Successful Adaptation to Climate Change across Scales." *Global Environmental Change* 15, no. 2 (July): 77–86. doi:10.1016/j.gloenvcha.2004.12.005.

Adhikari, Shubahm, Oliver Bellomi, and Alexander Mirescu. 2015. *Two Cities, One Storm: A Comparative Analysis of Post Sandy Resiliency Planning in Hoboken and Jersey City.* United Nations Office for Disaster Rick Reduction, The "State of DRR at the Local Level," New York, 2015. Accessed July 2, 2016. https://www.unisdr.org/we/inform/publications/48432.

Aldrich, Daniel P. 2012. *Building Resilience: Social Capital in Post-Disaster Recovery.* Chicago: University of Chicago Press.

American Society of Civil Engineers. 2013. "Infrastructure Report Card." Accessed August 20, 2016. https://www.infrastructurereportcard.org/making-the-grade/report-card-history/2013-report-card/.

Baechler, Gunther. 1998. Why Environmental Transformation Causes Violence: A Synthesis. *Environment Change and Security Project Report*, Washington, DC: Wilson Center (pp. 24–39). https://www.wilsoncenter.org/sites/default/files/ACF26C.pdf#page=24.

Barnet, Clive. 2005. "The Consolations of 'Neoliberalism." *Geoforum* 36, no. 1 (January): 7–12. doi:10.1016/j.geoforum.2004.08.006.

Barnett, J. 2003. "Security and Climate Change." *Global Environmental Change-Human and Policy Dimensions* 13, no. 7: 7–17, Accessed August 12, 2016. http://tyndall.ac.uk/publications/security-and-climate-change.

Barnett, Jon, and W. Neil Adger. 2007. "Climate Change, Human Security and Violent Conflict." *Political Geography* 26: 639–655. doi:10.1016/j.polgeo.2007.03.003.

Barry, Aandrew, Thomas Osborne, and Nickolas Rose, eds. 1996. *Foucault and Political Reason: Liberalism, Neoliberalism and Rationalities of Government*. Chicago, IL: University of Chicago Press.

Bataille, George. 1985. *Visions of Excess: Selected Writings, 1927–1939*. Trans. A. Stoekl, Carl R. Lovitt, and Donald M. Leslie Jr. Minneapolis: University of Minnesota Press.

Battistoni, Alyssa. 2013. "Back to No Future." *Jacobin: Reason in Revolt*. Accessed November 8, 2016. https://www.jacobinmag.com/2013/06/back-to-no-future.

Bennet, Jane. 2010. *Vibrant Matter: A Political Ecology of Things*. London, England: Duke University Press.

Benz, Arthur, Jorg Kemmerzell, Michel Knodt, and Anne Tews. 2015. "The Translocal Dimension of Local Climate Policy. Sustaining and Transforming Local Knowledge Orders through Trans-local Action in three German Cities." *Urban Research Practice* 8, no. 3 (May): 319–335. doi:10.1080/17535069.2015.1051380.

Berkes, Fikret, and Dyanna Jolly. 2001. "Adapting to Climate Change: Socio-ecological Resilience in a Canadian Western Arctic Community." *Conservation Ecology* 5, no. 2: 18. [online] URL: http://www.consecol.org/vol5/iss2/art18/.

Berkes, Fikret, and C Folke. 1998. *Linking Social and Ecological Systems: Management Practices and Social Mechanisms for Building Resilience*. New York: Cambridge University Press.

Betsill, Michele M. and Harriet Bulkeley. 2006. "Cities and the Multilevel Governance of Global Climate Change." *Global Governance: A Review of Multilateralism and International Organizations* 12, no. 2 (April): 141–159. 1075–2846.

Biermann, Frank. 2012. "The Overall Effects of Partnerships for Sustainable Development: More Smoke Than Fire?" In *Public-Private Partnerships for Sustainable Development: Emergence, Influence and Legitimacy,* edited by Philipp Pattberg and Biermann Frank, 288. Cheltenham, UK: Edward Elgar Publishing.

Biesbroek, G. Robert, Rob J. Swart, Timothy R. Carter, Caroline Cowan, Thomas Henrichs, Hanna Mela, Michael D. Morecroft, and Daniela Rey. 2010. "Europe Adapts to Climate Change: Comparing National Adaptation Strategies." *Global Environmental Change* 20, no. 3: 440–450. doi:10.1016/j.gloenvcha.2010.03.005.

Blachman-Biatch, Isadora, Keith Edgemon, Peter Hull, and Anna Taylor. 2013. "*The Resilient Social Network Occupy Sandy #SuperstormSandy*." Homeland Security Studies and Analysis Institute (HSSAI) for the Department of Homeland Security Science and Technology Directorate. http://newyork.resiliencesystem.org/department-homeland-security-resilient-social-network-occupysandy-superstormsandy; https://assets.documentcloud.org/documents/1357203/the-resilient-social-network.pdf.

Bogaert, Koenraad. 2013. "Contextualizing the Arab Revolts: The Politics Behind the Three Decades of Neoliberalism in the Arab World." *Middle East Critique* 22, no. 3: 213–234. doi:10.1080/19436149.2013.814945.

Brady, Michelle. 2016. "Neoliberalism, Governmental Assemblages, and the Ethnographic Imaginary," In *Governing Practices: Neoliberalism, Governmentality, and the Ethnographic Imaginary,* edited by Brady Michelle and Randy K. Lippert, 3–34. Toronto, Canada: University of Toronto Press.

Braun, Bruce. 2014. A New Urban Dispositif? Governing life in an Age of Climate Change. *Environment and Planning D: Society an Space*, 32, no. 32 (January): 49–64. http://doi.org/10.1068/d4313.

Brenner, Neil. 2004. *New State Spaces: Urban Governance and the Rescaling of Statehood.* Oxford, United Kingdom: Oxford University Press.

Brenner, Neil, Jamie Peck, and Nik Theodore. 2009. "Variegated Neoliberalization: Geographies, Modalities, Pathways." *Global Networks* 10, no. 2 (August): 182–222. doi:10.1111/j.1471-0374.2009.00277.x.

Brenner, Neil, and Nik Theodore. 2005. "Neoliberalism and the Urban Condition." *City* 9, no. 1 (October): 101–107. doi:10.1080/13604810500092106.

Brenner, Neil, and Nik Theodore. 2002. "Cities and the Geographies of 'Actually Existing Neoliberalism'." *Antipode* 34, no. 3 (July): 349–379. doi:10.1111/1467-8330.00246.

Briggs, Chad. 2017. "Foresight Tools and Early Warning Systems: Vulnerability Assessments for Abrupt and Non-Linear Climate Risks." In *Epicenter of Climate and Security: The New Geostrategic Landscape of the Anthropocene,* edited by Caitlin E. Werrell and Francesco Femia. Washington, DC: The Center for Climate and Security. 2017. Accessed September 29, 2017. https://climateandsecurity.org/reports/.

Brockling, Ulrich. 2016. *The Entrepreneurial Self: Fabricating a New Type of Subject.* London, United Kingdom: Sage Publications.

Broto, Vanessa Castán, and Harriet Bulkeley. 2013. "A Survey of Urban Climate Change Experiments in 100 Cities." *Global Environmental Change* 23, no. 1 (February): 92–102. doi:10.1016/j.gloenvcha.2012.07.005.

Brown, Wendy. 2005. *Edgework: Critical Essays on Knowledge and Politics.* Princeton, NJ: Princeton University Press.

Busby, Joshua. 2017. "Mapping Epicenters of Climate Security Vulnerabilities." In *Epicenter of Climate and Security: The New Geostrategic Landscape of the Anthropocene,* edited by Caitlin E. Werrell and Francesco Femia. Washington, DC: The Center for Climate and Security. Accessed September 29, 2017. https://climateandsecurity.org/reports/.

Buzan, Barry, Ole Waever, and Jaap De Wilde. 1998. *Security: A Framework for Analysis*. Boulder, CO: Lynne Rienner Publishers.
Carmines, Edward G, and James A. Stimson. 1993. "On the Evolution of Political Issues." In *Agenda Formation,* edited by William H. Riker, 151–191. Ann Arbor, MI: The University of Michigan Press.
Castree, Noel, and Bruce Braun, eds. 2001. *Social Nature: Theory, Practice, and Politics*. Malden, MA: Blackwell.
Cay, Johnston D. 2012. "Everyday Armageddon: Disaster Far Greater Than Sandy Looms - Unless We Move Fast to Fix a Broken System." *Newsweek,* December 3 2012. https://www.highbeam.com/doc/1G1-316794537.html.
Center for Climate and Energy Solutions. 2013. "Federal Congress." Accessed August 18, 2016. http://www.c2es.org/federal/congress/108/summary-mccain-lieberman-climate-stewardship-act-2003.
Center for Climate and Energy Solutions. 2013. "International Negotiations." Accessed August 10, 2016. http://www.c2es.org/international/negotiations/cop-19/summary.
Center for Climate and Energy Solutions. 2004. "International Negotiations." Accessed August 25, 2016. http://www.c2es.org/international/negotiations/cop-10/summary.
Center for Climate and Energy Solutions. 2015. "Outcomes of the U.N. Climate Change Conference in Paris." Accessed July 10, 2016. http://climateaction2020.unfccc.int/.
Chakrabarty, Dipesh. 2012. "Postcolonial Studies and the Challenge of Climate Change." *New Literary History* 43, no. 1 (Winter): 1–18. doi:10.1353/nlh.2012.0007.
City of New York, The. 2013a. "PLANY Progress Report: A Greener, Greater New York." *Mayor Michael Bloomberg*. http://smedia.nyc.gov/agencies/planyc2030/pdf/planyc_progress_report_2013.pdf.
City of New York, The. 2013b. "PlanNYC: A Stronger, More Resilient New York." Accessed July 28, 2016. http://s-media.nyc.gov/agencies/sirr/SIRR_singles_Lo_res.pdf.
Clarke, Nick. 2009. "In What Sense 'Spaces of Neoliberalism'? The New Localism, the New Politics of Scale, and Town Twinning." *Political Geography* 28, no. 8 (November): 496–507. doi:10.1016/j.polgeo.2009.12.001.
Climate Institute, The. 2010. "The Cancun Agreement: A Preliminary Assessment: Policy Brief." Accessed August 23, 2016. http://www.climateinstitute.org.au/verve/_resource/tci_thecancunagreement_policybrief.pdf.
Coaffee, Jon. 2009. *Terrorism, Risk, and the Global City: Towards Urban Resilience*. New York: Routledge.
Collier, Stephen J. 2012. "Neliberalism as big Leviathan, or . . . ? A response to Wacquant and Hilgers." *Social Anthropology* 20, no. 2 (May): 186–195. doi:10.1111/j.1469-8676.2012.00195.x.
Congleton, Roger D. 2006. "The Story of Katrina: New Orleans and the Political Economy of Catastrophe." *Public Choice* 127, no. 1/2 (April): 5–30.
Consolidated of New York Con Edison. 2013. "Annual Report, Con Edison." *Everything Matters*. Accessed June 1, 2017. http://phx.corporate-ir.net/phoenix.zhtml?c=61493&p=irol-reportsannual.

Convention on Biological Diversity. 2014. "*Report on the Twelfth Meeting of the Conference of the Parties to the Convention on Biological Diversity.*" Pyeongchang, Republic of Korea, October 6–17, 2014. Accessed November 11, 2016. https://www.cbd.int/cop12/.

Constanza, Robert. 2017. "Claim the Sky." In *Building a Climate Resilient Economy and Society: Challenges and Opportunities,* edited by K. N. Ninan and Makoto Inoue, 279–281. Northhamptom, MA: Edward Elgar Publishing.

Corbridge, Stuart. 1998. "Beneath the Pavement Only Soil": The Poverty of Post-Development." *Journal of Development Studies* 346, no. 6 (February): 138–148. 10.1080/00220389808422549.

Corfee-Morlot, Jan, Lamia Kamal-Chaoui, Michael G. Donovan, Ian, Cochran, and Alexis Robert. 2009. "Cities, Climate Change and Multilevel Governance." OECD Environment Working Papers, no. 14. OECD Publishing. doi:10.1787/220062444715.

Crutzen, Paul, and Eugene Stoermer. 2000. "The Anthropocene." *The Global Change Newsletter, The International Geosphere-Biosphere Programme (IGBP).* 2000. Accessed September 22, 2016. http://www.igbp.net/globalchange/anthropocene.4.1b8ae20512db692f2a680009238.html.

Dabelko, Geoffrey. 2006. "From Threat to Opportunity: Exploiting Environmental Pathways to Peace. Prepared For "Environmental, Peace and the Dialogue among Civilizations and Cultures" May 2006, Tehran, Islamic Republic of Iran." Accessed August 14, 2016. http://www.wilsoncenter.org/sites/default/files/pathways.pdf.

Dalby, Simon. 2013. "Rethinking Geopolitics. Climate Security in the Anthropocene." *Global Policy* 5, no. 1 (February): 1–9. doi:10.1111/1758-5899.12074.

Dalby, Simon. 2007. "Anthropocene Geopolitics. Globalization, Empire, Environment, and Critique." *Geography Compass* 1, no. 1 (January): 103–118. doi:10.1111/j.1749-8198.2007.00007.x.

Davies, Jonathan. 2014. "Rethinking Urban Power and the Local State: Hegemony, Domination and Resistance in Neoliberal Cities." *Urban Studies* 51, no. 15: 3215–3232. doi:10.1177/0042098013505158.

Davis, Mike. 2010. "Who will build the ark?" *New Left Review* 61 (January): 29–46. https://ecology-and-socialism.wikispaces.com/file/view/NLR+-+Who+Will+Build+the+Ark+Mike+Davis.pdf/173877777/NLR+-+Who+Will+Build+the+Ark+Mike+Davis.pdf.

Devcka-Rinear, A., J., Limbacher, E., Marshall, L. R., Murray, and M, Ochsner. 2017. "*The Long Road Home: Understanding Sandy Recovery and Lessons for Future Storms Five Years Later.*" New Jersey: New Jersey Resource Project. 2017. Accessed October 28, 2017, http://www.newjerseyrp.org/assets/long-road-home-101717.pdf.

Dimmer, Christian, and Jan Lindberg. 2014. "Mapping Social Innovation and Strengthening Community: Bottom-up Recovery Initiatives and Community Spaces in Post-Disaster Ishinomaki, Japan." Paper presented at the *International Alliance for Sustainable Urbanization and Regeneration Conference, October 2014.* https://www.researchgate.net/publication/303878372_Mapping_Social_Innovation_and_Strengthening_Community_Resilience_Bottom-up_Re-_covery_Initiatives_and_Community_Spaces_in_Post-Disaster_Ishinomaki_Japan.

Dinan, Terry. 2016. "Potential Increases in Hurricane Damage in the United States: Implications for the Federal Budget." Congress of the United States Congressional Budget Office. Accessed October 19, 2016. https://www.cbo.gov/sites/default/files/114th-congress-2015-2016/reports/51518-hurricane-damage-onecol.pdf.

DuBois, Bryce B. 2016. "Beaches, People and Change: A Political Ecology of Rockaway Beach after Hurricane Sandy." Ph.D. diss., Graduate Center, City University of New York.

Dyson, Michael Eric. 2006. *Come Hell or High Water: Hurricane Katrina and the Color of Disaster.* New York: Civitas Books.

Ekberg, Merryn. 2007. "The Parameters of the Risk Society: A Review and Exploration." *Current Sociology* 55, no. 3 (2007): 343–366. doi:10.1177/0011392107076080.

England, Kim, and Kevin Ward. 2007. *Neoliberalization: States, Networks, Peoples.* Malden, MA: Blackwell.

Environmental Change and Security Program. 2016. "Environmental Cooperation as a Peace Building Tool: Security Brief and Webchat." Accessed August 31, 2016. http://wilsoncenter.org/article/promoting-environmental-cooperation-peace-building-tool-security-brief-and-webchat.

Federal Emergency Management Agency. 2013. "The Stafford Act: Robert T. Stafford Disaster Relief and Emergency Assistance Act, as Amended." Accessed September 29, 2016. http://www.Fema.gov/media-library-data/138315366995521f970b19e8aaa67087b7da 9f4af706e/stafford_act_booklet_042213_508e.pdf.

Federal Emergency Management Agency. 2011. "A Whole Community Approach to Emergency Management: Principles, Themes, and Pathways for Action." Accessed July 2, 2014. https://www.fema.gov/media-library-data/20130726-1813-25045-3330/whole_community_dec2011__2_.pdf.

Femia, Francesco, and Caitlin E. Werrel. 2017."Climate Change, the Erosion of State Sovereignty, and New World Order." In *Epicenter of Climate and Security: The New Geostrategic Landscape of the Anthropocene,* edited by Caitlin E. Werrell and Francesco Femia. Washington, DC: The Center for Climate and Security. Accessed September 19, 2017. https://climateandsecurity.org/reports/.

Ferguson, James, and Akhil Gupta. 2002. "Spatializing States: Toward an Ethnography of Neoliberal Governmentality." *American Ethnologist* 29, no. 4 (November). doi:10.1525/ae.2002.29.4.981.

Foucault, Michel. 2008. *The Birth of Biopolitics: Lectures at the College De France 1978–1979.* New York: Palgrave MacMillan, 2008.

Foucault, Michel. 1997. "Of Other Spaces: Utopias and Heterotopias." In *Rethinking Architecture: A Reader in Cultural History*, edited by Neil Leach, 330–336. New York: Routledge.

Foucault, Michel. 1994. "The Birth of Biopolitics." In *Foucault: Ethics Subjectivity and Truth, Vol 1,* edited by Paul Rainbow. New York: The New Press.

Foucault, Michel. 1991. "Governmentality." In *The Foucault Effect: Studies in Governmentality*, edited by Graham Burchell, Colin Gordon and Peter Miller, 87–104, Chicago, IL: Chicago University Press.

Freeman, Andrew. 2013. "U. S. Dominated global disaster losses in 2012: Swiss Re. *Climate Central,* Accessed July 2, 2017. http://www.climatecentral.org/news/us-dominated-global-disaster-losses-in-2012-insurer-reports-15814.

Feuer, Alan. 2012. "Occupy Sandy: A Movement Moves to Relief." *The New York Times*, November 9, 2012. Accessed July 10, 2016, http://www.nytimes.com/2012/11/11/nyregion/where-fema-fell-short-occupy-sandy-was-there.html.

Giddens, Anthony. 2011. *The Politics of Climate Change*. Cambridge, United Kingdom: Polity Press.

Giroux, Henry A. 2006. *Stormy Weather: Katrina and the Politics of Disposability*. Boulder, CO: Paradigm Publishers.

Global Covenant of Mayors for Climate Change. 2016. Accessed August 31, 2016. http://www.compactofmayors.org/.

Goldstein, Adam H., and Constantine Samaras. 2017. "Dire Straits: Strategically Significant International Waterways in a Warming World." In *Epicenter of Climate and Security: The New Geostrategic Landscape of the Anthropocene,* edited by Caitlin E. Werrell and Francesco Femia. Washington, DC: The Center for Climate and Security. Accessed October 15, 2017. https://climateandsecurity.org/reports/.

Gonzales, Carmen, Alice Kaswan, Robert Verchick, Yee Huang, Shawn Bowen, and Nowal Jamhour. 2016. "Climate Change, Resilience, and Fairness: How Nonstructural Adaptation Can Protect and Empower Socially Vulnerable Communities on the Gulf Coast." *Center for Progressive Reform,* Accessed July 7, 2017. http://progressivereform.org/articles/Climate_Change_Resilience_Gulf_Coast_1603.pdf.

Goodman, Sherri. 2012. "What Is Environmental Security?" *Yale Insights. Yale School of Management.* April 15, 2012, Accessed June 12, 2017. http://insights.som.yale.edu/insights/what-is-environmental-security.

Graham, S. 1995. "From Urban Competition to Urban Collaboration? The Development of Interurban Telematics Networks." *Environment & Planning C: Government & Policy* 13, no. 4 (December): 503–524. doi:10.1068/c130503.

Grothmann, Torsten, and Anthony Patt. 2005. "Adaptive Capacity and Human Cognition: The Process of Individual Adaptation to Climate Change." *Global Environmental Change Part A: Human & Policy Dimensions* 15 no. 3 (October): 199–213. doi:10.1016/j.gloenvcha.2005.01.002.

Gunderson, Lance H. and Crawford Stanley Holling, eds. 2002. *Panarchy: Understanding Transformations in Human and Natural Systems.* Washington, DC: Island Press.

Gunderson, Lance H., Craig R. Allen, and C. S. Holling, eds. 2010. *Foundations of Ecological Resilience*. Washington, DC: Island Press.

Gupta, Nayanee, Christopher T. Clavin, Zoe E. Petropoulos, Austin B. Mudd, Rashida Nek, and Sally S. Tinkle. 2016. *Case Studies of Community Resilience Policy.* IDA Science and Technology Policy Institute. 2016. MD: National Institute of Standards and Technology, U. S. Department of Commerce Engineering Laboratory. Accessed July 20, 2016. doi:10.6028/NIST.GCR.16–002.

Haddow, George D., and Jane A. Bullock. 2003. *Introduction to Emergency Management*. Boston, MA: Butterworth-Heinemann.

Haraguchi, Masahiko, and Soojun Kim. 2014. *Critical Infrastructure Systems: A case study of the interconnectedness of risks posed by Hurricane Sandy for New York City.* United Nation Office for Disaster Risk Reduction. "Global Assessment Report on Disaster Risk Reduction." Accessed September 8, 2016. http://www.preventionweb.net/publications/view/48347.

Haraway, Donna. 2016. "Staying with the Trouble: Anthropocene, Capitalocene, Chthulucene." In *Anthropocene or Capitalocene: Nature, History, and the Crisis of Capitalism*, edited by Jason W. Moore, 34–76. Oakland, CA: PM Press.

Harvard Kennedy School. 2012. "The Durban Platform Negotiations: Goals and Options." Harvard Project on Climate Agreements, and Belfer Center for Science and International Affairs. Accessed August 28, 2016. http://belfercenter.ksg.harvard.edu/publication/22196/durban_platform_negotiations.html.

Harvey, David. 2007. *A Brief History of Neoliberalism*. New York: Oxford University Press.

Harvey, David. 2003. *The New Imperialism*. New York: Oxford University Press.

Harvey, David. 1992. *The Condition of Postmodernity*. Oxford, United Kingdom: Basil Blackwell Ltd.

Hawkins, Christopher V., Rachel M. Krause, Richard C. Feiock, and Call Curley. 2015. "Making Meaningful Comments: Accounting for Variation in Cities' Investments of Staff and Fiscal Resources to Sustainability." *Urban Studies* 53, no. 39 (April): 1902–1924. doi:10.1177/0042098015580898.

Holling, Crawford Stanley. 1973. "Resilience and Stability of Ecological Systems." *Annual Review of Ecology and Systematics* 41, no. 23 (November): 1–23. doi:.04.110173.000245.

Holmberg, Karen. 2007. *Beyond the Catastrophe: The Volcanic Landscape of Baru, West Panama. In Living under the Shadow: Cultural Impacts of Volcanic Eruptions*. Walnut Creek, CA: Left Coast Press.

Homsy, George C., and Mildred E. Warner. 2015. "Cities and Sustainability: Polycentric Action and Multilevel Governance." *Urban Affairs Review* 51, no. 1 (April): 46–73. doi:10.1177/1078087414530545.

Horn, E. 2013. *Comprehensive Coastal Communities Playbook*. Long Island: Operation Resilient Long Island.

Horton, R., C. Little, V. Gornitz, D. A. Bader, and M. Oppenheimer. 2015. "New York City Panel on Climate Change 2015 Report: Sea Level Rise and Coastal Storms." *Annals of the New York Academy of Sciences*. 1336, no. 36–44. Building the Knowledge base for Climate Resiliency. July 26, 2016. doi:10.1111/nyas.12593.

Hsu, Angel, Andrew S. Moffat, Amy J. Weintfurter, and Jason D. Schwartz. 2015. "Toward a New Climate Diplomacy." *Nature Climate Change* 5 (May): 501–503.

Hughes, Sarah, and Andrea Sarzynski. 2015. "Building Capacity for Climate Change Adaptation in Urban Areas: Editor's Introduction." *Urban Climate* 14, no. 1 (December): 1–3. doi:10.1038/nclimate2594.

Jaleel, Rana. 2012. Into the storm: Occupy Sandy and the new sociality of debt. *Social Text: Periscope*. Accessed June 1, 2017. https://what-democracy-looks-like.org/into-the-storm-occupy-sandy-and-the-new-sociality-of-debt/.

Jenkins, Pamela. 2016. "A Sense of Place at Risk: Perspectives of Residents of Coastal Louisiana on Nonstructural Risk Reduction Strategies." *OXFAM America*, Accessed July 10, 2017. https://www.oxfamamerica.org/explore/research-publications/a-sense-of-place-at-risk/.

Jessop, Bob. 2007. "New Labor or the Normalization of Neoliberalism." *British Politics* 2, no. 2: 282–288. doi:10.1057/palgrave/bp.4200059.

Jimenez Dominguez, Bernardo. 2009. "Duality, Globalization, and Urban Culture: Guadalajara, Mexico," In *On Global Grounds: Urban Change and Globalization,* edited by Julia Nevarez and Gabriel Moser, 85–96. New York: Nova Science Publishers, Inc.

Johnson, Elizabeth, and Harlan Morehouse. 2014. "After the Anthropocene: Politics and Geographic Inquiry for a New Epoch." *Progress in Human Geography* 38, no. 3 (May): 439–456. doi:10.1177/0309132513517065.

Johnston, David Cay. 2012. "Everyday Armageddon." *Newsweek,* December 3, 2012. Accessed June 14, 2016. https://www.highbeam.com/doc/1G1-316794537.html.

John R. Justus, and Susan R. Fletcher. 2003. "Global Climate Change. Resources, Science, and Industry Division. CRS Issue Brief for Congress." CRS Issue Brief for Congress. Accessed November 11, 2016. http://xmanager.portalecnel.it/application/xmanager/projects/cnel/attachments/shadow_documentazioni_attachment/file_allegatos/000/081/740/il_20cambiamento_20del_20clima_20mondiale.pdf.

Kemmerzell, Jorg. 2016. "Trans-Local Action and Local Innovations in Climate Change Policy. A Comparative Analysis of German Major Cities." *Urban Affairs Association Conference Presentation.* 2016.

Keohane, Robert O., and David G. Victor. 2011. "The Regime Complex for Climate Change." *Perspectives on Politics* 9, no. 1 (March): 7–23. doi:10.1017/S1537592710004068.

Kern, Kristine, and Harriet Bulkeley. 2009. "Cities, Europeanization and Multi-Level Governance: Governing Climate Change through Transnational Municipal Networks." *Journal of Common Market Studies* 47, no. 2 (January): 309–332. doi:10.1111/j.1468-5965.2009.00806.x.

Kettle, Nathan P., and Kirstin Dow. 2016. "The Role of Perceived Risk, Uncertainty, and Trust on Coastal Climate Change Adaptation Planning." *Environment & Behavior* 48, no. 4 (September): 579–606. doi:10.1177/0013916514551049.

Kingdon, John W. 1995. *Agendas, Alternatives, and Public Policies.* 2nd ed. New York: Harper Collins College Publishers.

Klare, Michael T. 2001. *Resource Wars: The New Landscape of Global Conflict.* New York: Henry Holt and Company, A Metropolitan Book.

Klein, Richard J. T., Robert J. Nicholls, and Frank Thomalla. 2004. "Resilience to Natural Hazards: How Useful Is This Concept?" *Global Environmental Change Part B: Environmental Hazards* 5, no. 1–2: 35–45. doi:10.1016/j.hazards.2004.02.001.

Krauss, Celene. 1989. "Community struggles and the shaping of democratic consciousness." *Sociological Forum* 4, no. 2 (June): 227–239. doi:10.1007/BF01112422.

Krause, Rachel. 2013. "The Motivations Behind Municipal Climate Engagement." *Cityscape* 15, no. 1: 125–141. http://www.jstor.org/stable/41958960.

Krause, Rachel. 2011. "Symbolic or Substantive Policy? Measuring the Extent of Local Commitment to Climate Protection." *Environment and Planning C: Government and Policy* 29, no. 1 (January): 46–62. doi:10.1068/c09185.

Kroll-Smith, J. Stephen, and Stephen R. Couch. 1993. "Technological Hazards: Social Responses as Traumatic Stressors." In *International Handbook of Traumatic*

Stress Syndromes, edited by John P. Wilson and Beverly Raphael, 79–91. New York: Plenum Press, 1993. doi:10.1007/978-1-4615-2820-3_7.

Langer, Sidney. 2012. *From Slavery to 9/11: Readings in the Sociology and Social Psychology of Extreme Situations*. London: Taylor and Francis.

Larner, Wendy. 2003. "Neoliberalism." Guest Editorial. *Environment & Planning D: Society & Space* 21, no. 5 (October) 509–512.

Lefebrve, Henri. 2004. *Rhythmanalysis: Space, Time, and Everyday Life*. New York: Continuum Press.

Leitner, H., C. Pavlik, and E. Sheppard. 2002. "Networks, Governance, and the Politics of Scale: Inter-Urban Networks, and the European Union." *Geographies of Power: Placing Scale*. Oxford, UK: Blackwell.

Leitner, H. and E. Sheppard. 2002. "The City Is Dead, Long Live the Net": Harnessing European Interurban Networks for a Neoliberal Agenda." *Antipode* 34, no. 3 (May): 495–518.

Lemke, Thomas. 2002. "Foucault, Governmentality, and Critique." *Rethinking Marxism: A Journal of Economic, Culture and Society* 14, no. 3: 49–64. doi:10.1080/089356902101242288.

Lin, Nan. 2001. *Social Capital: A Theory of Social Structure and Action*. Cambridge: Cambridge University Press.

Liu, Amy, Roalnd Anglin, Richard Mizelle, Allison Plyer. 2011. *Resilience and Opportunity: Lessons from the U. S. Gulf Coast after Katrina and Rita*. Washington, DC: Brookings Institution Press.

Lorey, Isabell. 2015. *State of Insecurity: Government of the Precarious*. London, UK: Verso Publishers.

Mairie de Paris. 2017. "Strategie De Resilience De Paris: Fluctuat Nec Mergitur." *100resilient cities*. Accessed September 15, 2017. http://www.100resilientcities.org/strategies/paris/.

Martinez, Elizabeth, and Arnoldo Garcia. 2000. "What Is Neoliberalism? A Brief Definition of Activists." *CorpWatch*. http://www.corpwatch.org/article.php?id=376.

Mason, Susan G., and Michail Fragkias. 2016. "Mental Models and the Robustness of Institution: Do They Influence Metropolitan Planning Organization Involvement in Climate Change Policy?" *Urban Affairs Association Conference 2016*.

Mbembe, J. Achille, 2003. "Necropolitics." Translated by Libby Meintjes. *Public Culture* 15, no. 1 (Winter): 11–40. doi:10.1215/08992363-15-1-11.

McCann, Eugene A. 2008. "Expertise, Truth, and Urban Mobilities: Global Circuits of Knowledge and the Development of Vancouver, Canada's 'Four Pillar' Drug Strategy." *Environment and Planning A* 40, no. 4 (January): 885–904. doi:10.1068/a38456.

McLeman, Robert. 2017. "Migration and Displacement in a Changing Climate." In *Epicenter of Climate and Security: The New Geostrategic Landscape of the Anthropocene*, edited by Caitlin E. Werrell and Francesco Femia. Washington, DC: The Center for Climate and Security. Accessed August 27, 2017. https://climateandsecurity.org/reports/.

McKensie, W. 2015. *Molecular Red: Theory for the Anthropocene*. New York: Verso.

Meakin, Stephanie. 1992. "The Rio Earth Summit: Summary of the United Nations Conference on Environment and Development." Accessed October 31, 2016. http://publications.gc.ca/collection-R/LoPBdP/BP/bp317-e.htm.

Micheletti, Michele, ed. 2003. *Politics of Virtue and Shopping: Individuals, Consumerism, and Collective Action.* London, United Kingdom: Palgrave.

Miller, Demond Shondell, and Jason David Rivera. 2008. *Hurricane Katrina and the Redefinition of Landscape*. Lanham, MD: Lexington Books.

Milman, Oliver. 2017. "Hurricane Sandy Five Years Later: No One Was Ready for What Happened After." *The Guardian* [New York] Accessed October 28, 2017 https://www.theguardian.com/us-news/2017/oct/27/hurricane-sandy-five-years-later-climate-change.

Mirzoef, Nicholas. 2014. "Visualizing the Anthropocene." *Public Culture* 26, no. 2: 213–232. doi:10.1215/08992363-2392039.

Moore, Jason W. 2016. *Anthropocene or Capitalocene?: Nature, History, and the Crisis of Capitalism.* Oakland, CA: PM Press.

Mourad, Bessma, and Amy Luers. 2017. "Tools for Understanding Systematic Risks Like Climate Change." In *Epicenter of Climate and Security: The New Geostrategic Landscape of the Anthropocene,* edited by Caitlin E. Werrell and Francesco Femia. Washington, DC: The Center for Climate and Security. Accessed October 28, 2017. https://climateandsecurity.org/reports/.

Murakami, Kayo, David Murakami Wood, Hiroshi Tomita, Satoshi Miyake, Rieko Shiraki, Kayo Murakami, Koji Itonaga, and Christian Dimmer. 2014. "Evolving Place Governance Innovation and Pluralizing Reconstruction Practices in Post-Disaster Japan." *Planning, Theory and Practice* 15, no. 2 (May): 260–265. doi:10.1080/14649357.2014.902909.

Murray, Patrick. 2017. "NJ Sandy Panel: Still Dissatisfied with Recovery: Despite Improvement, Many Survivors still Face Hurdles 5 Years Later." October 25, 2017. Accessed October 28, 2017. https://www.monmouth.edu/polling-institute/reports/MonmouthPoll_NJSANDY_102517/.

New York Department of Environmental Protection Climate Change Program, The. 2008. "*Assessment and Action Plan: A Report Based on the Ongoing Work of the Department of Environmental Protection, Climate Change Task Force. Emily Lloyd, Commissioner.*" Accessed February 15, 2017. http://www.nyc.gov/html/dep/pdf/climate/climate_complete.pdf.

Ng, Edward, and Chao Ren. 2015. *The Urban Climatic Map: A Methodology for Sustainable Urban Planning.* New York: Routledge.

Ninan, K. N., and Makoto Inoue, eds. 2017. *Building Climate Resilient Economy and Society: Challenges and Opportunities.* Northampton, MA: Edward Elgar Publishing Limited.

Nolan, Bruce. 2011. "6 Years Later, Hurricane Katrina's Scars Linger Alongside Robust Recovery." *The Times-Picayune, Greater New Orleans,* August 28, 2011. Accessed November 13, 2013. http://www.nola.com/katrina/index.ssf/2011/08/6_years_later_hurricane_katrin.html.

Norris, Fran H., Susan P. Steven, Betty Pfefferbaum, Karen F. Wyche, and Rose L. Pfefferbaum. 2008. "Community resilience as a metaphor, theory, set of capacities, and strategy for disaster readiness." *American Journal of Community Psychology* 41, no. 1–2 (March): 127–150. doi:10.1007/s10464-007-9156-6.

NYS 2100 Commission. 2013. *Recommendations to Improve the Strength and Resilience of the Empire State's Infrastructure.* New York: State, Governor Cuomo. Accessed July 26, 2016. http://www.lcbp.org/wp-content/uploads/2013/03/NYS2100.pdf.

O'Callahan, Ted. 2012. What is environmental security? Interview to Sherri Goodman. *Yale Insights.* Yale School of Management. April 15, 2012. Accessed November 1, 2016. http://insights.som.yale.edu/insights/what-is-environmental-security.

Oliver-Smith, Anthony. 2001. "Theorizing Disasters: Nature, Power, and Culture." In *Catastrophe and Culture: The Anthropology of Disaster,* edited by Susanna M. Hoffman and Anthony, Oliver-Smith, 23–48. Santa Fe, NM: School of American Research Press.

Palsson, Gisli, and Heather Anne Swanson. 2016. "Down to Earth: Geosocialities and Geopolitics." *Environmental Humanities* 8, no. 2 (November): 149–171. doi:10.1215/22011919-3664202.

Parenti, Christian. 2016. "Environment-Making in the Capitalocene: Policy Ecology of the State." In *Anthropocene or Capitalocene: Nature, History, and the Crisis of Capitalism,* edited by Jason Moore, 166–184. Oakland, CA: PM Press.

Peck, Jamie. 2002. "Political Economies of Scale: Fast Policy, Interscalar Relations, and Neoliberal Workfare." *Economic Geography* 78, no. 3 (July): 331. doi:10.1111/j.1944-8287.2002.tb00190.x.

Peck, Jamie. 2001. "Neoliberalizing States: Thin Policies/Hard Outcomes." *Progress in Human Geography* 25, no. 3 (September): 445–455. doi:10.1191/03091320 1680191772.

Peck, Jamie, and Adam Tickell. 2002. "Neoliberalizing Space." *Antipode* 34, no. 3 (July): 380–404. doi:10.1111/1467-8330.00247.

Peck, Jamie, and Nik Theodore. 2001. "Exporting Workfare/Importing Welfare-to-Work: Exploring the Politics of Third Way Policy Transfer." *Political Geography* 20, no. 4 (May): 427–460. doi:10.1016/S0962-6298(00)00069-X.

Peck, Jamie, Nik Theodore, and Neil Brenner. 2009. "Neoliberal Urbanism: Models, Moments, Mutation." *SAIS Review of International Affairs* 29, no. 1 (January): 49–66. 10.1353/sais.0.0028.

Peluso, Nancy Lee, and Michael Watts. 2001. *Violent Environments.* Ithaca, NY: Cornell University Press.

Perry, David M. 2017. "There Are No Natural Disasters: A Conversation with James Remes." *Pacific Standard,* October 4, 2017. Accessed October 5, 2017. https://psmag.com/economics/there-are-no-natural-disasters.

Pew Center on Global Climate Change. 2009. *Fifteenth Session of the Conference of the Parties to the United Nations Framework Convention on Climate Change and Fifth session of the meeting of the parties to the Kyoto Protocol.* December 7–18, 2009. Accessed 23 August 2016. http://www.c2es.org/docUploads/copenhagen-cop15-summary.pdf.

Picou, J. Steven, Brent K. Marshall, and Duane A. Gill. 2004. "Disaster, Litigation, and the Corrosive Community." *Social Forces* 82, no. 4 (June): 1493–1522. doi:10.1353/sof.2004.0091.

Platt, Rutherford. 1999. *Disasters and Democracy: The Politics of Extreme Natural Events*, Washington, DC: Island Press.

Plyer, Allison, Nihal Shrinath, and Mack Vicki. 2015. "The New Orleans Index at Ten: Measuring Greater New Orleans Progress Towards Prosperity." *The Data Center and the Brookings Institution Metropolitan Policy Program.* https://www.datacenterresearch.org/reports_analysis/new-orleans-index-at-ten/.

Purcell, Mark Hamilton. 2008. *Recapturing Democracy: Neoliberalization and the Struggle for Alternative Urban Futures.* New York: Routledge.

Purdy, Jedediah. 2015. *After Nature: A Politics for the Anthropocene.* Cambridge, MA: Harvard University Press.

Ritchie, Liesel Ashley, and Duane A. Gill. 2007. "Social Capital Theory as an Integrating Theoretical Framework in Technological Disaster Research." *Sociological Spectrum* 27, no. 1 (December): 103–129. doi:10.1080/02732170601001037.

Rosenzweig, Cynthia, and William Solecki. 2015. "New York City Panel on Climate Change 2015 Report." *Annals of the New York Academy of Sciences.* Building the Knowledge base for Climate Resiliency. Accessed July 26, 2016. http://onlinelibrary.wiley.com/doi/10.1111/nyas.12593/epdf.

Ruby, Andreas. 1998. Interview to Paul Virilio, "Surfing the Accident." *The Art of the Accident.* http://v2.nl/archive/articles/surfing-the-accident.

Saenz, R. 2005. *Beyond New Orleans: The Social and Economic Isolation of Urban African Americans.* Population Reference Bureau. Accessed October 26, 2005. http://www.prb.org/template.cfm?section=PRB&template=content/contentgroups/05_Articles/The_Social_And_Economic_American.htm.

Shaviro, Steven. 2015. *No Speed Limit: Three Essays on Accelerationism.* Minnesota, MN: Minnesota University Press.

Saloojee, Ozayr, and Vincent DeBritto. 2014. "Visible and Invisible Infrastructures: Alternative Futures in Resiliency, Failure, and Design." *ARCC/EAAE, Beyond Architecture: New Intersections and Connections.* Accessed July 10, 2013. http://www.arcc-journal.org/index.php/repository/article/view/295.

Shi, Linda, Eric Chu, Isabelle Anguelovski, Alexander Aylett, Jessica Debats, Kian Goh, Todd Schenk, Karen C. Seto, David Dodman, Debra Roberts, J. Timmons Roberts, and Stacy D. VanDeveer. 2016. "Roadmap Towards Justice in Urban Climate Adaptation Research." *Nature Climate Change* 6, no. 2 (July): 131–137. doi:10.1038/nclimate2841.

Simonet, Guillaume, and Sandra Fatoric. 2016, "Does 'Adaptation to Climate Change' Mean Resignation or Opportunity?" *Regional Environmental Change* 16, no. 3 (March): 789–799.

Slaughter, Anne Marie. 2004. *A New World Order.* Princeton, NJ: Princeton University Press.

Smith, Adrian, Alison Stenning, and Katie Willis, eds. 2008. *Social Justice and Neoliberalism: Global Perspectives.* London, UK: Zed Books.

Sobel, Russell S., and Peter T. Leeson. 2006. "Government's Response to Hurricane Katrina: A Public Choice Analysis." *Public Choice* 127, no. 1–2 (April): 55–73. http://jstor.org/stable/30026772.

Sparke, Matthew. 2005. *In the Space of Theory.* Minneapolis, MN: University of Minnesota Press.

Spehr, Christopher. 2007. "Free Cooperation." In *The Art of Free Cooperation*, edited by Trevor Scholz, and Geert Lovink. New York: Autonomedia, Institute for Distributed Creativity, Institute of Network Cultures.

Sprague, Teresa. 2016. "Challenging Local Traditions in a Changing Environment: Impacts of Local Identities on Planning for Disaster Risk Management in the Sub-Carpathians." *Urban Affairs Association,* Paper Presentation.

Springer, Simon. 2015. "Postneoliberalism?" *Review of Radical Political Economics* 47, no. 1 (February): 5–17. doi:10.1177/0486613413518724.

Springer, Simon. 2010. "Neoliberalism and Geography: Expansions, Variegations, Formations." *Geography Compas* 4, no. 8 (August): 1025–1038. doi:10.1111/j.1749-8198.2010.00358.x.

Springer, Simon, Kean Birch, and Julie MacLeavy. 2016. "An Introduction to Neoliberalism." In *The Handbook of Neoliberalism*, edited by Simon Springer, Kean Birch and Julie MacLeavy. London, UK: Routledge.

State of the Union Address, President Obama. 2011. Remarks of President Obama in State of the Union Address. Januray 25, 2011. Accessed July 18, 2016. https://obamawhitehouse.archives.gov/the-press-office/2011/01/25/remarks-president-barack-obama-state-union-address-prepared-delivery.

Swyngedouw, Erik. 2007. "The Post-Political City." In *Urban Politics Now. Re-Imagining Democracy in the Neoliberal City*, 58–77. Rotterdam: NAI Publishers.

Swyngedouw, Erik, Frank Moulaert, and Arntxa Rodriguez. 2002. "Neoliberal Urbanization in Europe: Large-Scale Urban Development Projects and the New Urban Policy." *Antipode* 34, no. 3 (July): 542–577. doi:10.1111/1467-8330.00254.

Swyngedouw, Erik. 1989. "The Heart of the Place: The Resurrection of Locality in an Age of Hyperspace." *Geografiska Annaler, Series B, Human Geography* 71, no. 1: 31–42. doi:10.2307/490505.

Sylves, R. T. 2006. "President Bush and Hurricane Katrina: A Presidential Leadership Study." *The Annals of the American Academy of Political and Social Science* 604, no. 1 (March): 26–56. doi:10.1177/0002716205286066.

Thomas, Julia Adeney. 2014."History and Biology in Anthropocene: Problems of Scale, Problems of Values." *American Historical Review* 119, no. 5 (December): 1587–1607. doi:10.1093/ahr/119.5.1587.

Thrift, Nigel. 2005. *Knowing Capitalism.* London, England: Sage.

Tierney, Kathleen J. 2005. *Recent Developments in Us Homeland Security Policies and Their Impacts for the Management of Extreme Events.* Boulder, CO: University of Colorado at Boulder.

Tompkins, Emma L., and Neil W. Adger. 2004. "Does Adaptive Management of Natural Resources Enhance Resilience to Climate Change?" *Ecology & Society* 9, no. 2 (October): 10. http://www.ecologyandsociety.org/vol9/iss2/art10.

Tsing, Anna. 2000. "The Global Situation." *Cultural Anthropology* 15, no. 2 (August): 137–163. doi:10.1525/can.2000.15.3.327.

United States Department of Homeland Security. 2013. "National Response Framework." 2nd Edition. Accessed January 25, 2016. https://www.fema.gov/media-library-data/20130726-1914-25045-1246/final_national_response_framework_20130501.pdf.

United States Department of Housing and Urban Development. 2013. "Hurricane Rebuilding Strategy: Stronger Communities, A Resilient Region." Hurricane Sandy Rebuilding Task Force. Accessed October 24, 2016. https://www.hud.gov/sites/documents/HSREBUILDINGSTRATEGY.PDF.

United Nations Office for Disaster Risk Reduction, The. 2014. "Economic and Human Capital Impacts of Disasters in the Past 10 Years." Accessed July 24, 2017. http://www.unisdr.org/files/42862_economichumanimpact20052014unisdr.pdf.

United Nations Office for Disaster Risk Reduction, The 2013. "Disaster Impacts/2000-2012." Accessed July 7, 2016. http://www.unisdr.org.

United Nations Office for Disaster Risk Reduction, The. 2011. "Climate Related Disaster 1980–2011, Annual Report." Accessed March 14, 2016. http://www.unisdr.org/files/27627_ar2011v2.pdf.

United Nations Office for the Coordination of Humanitarian Affairs (OCHA). 2011. "Annual Report." Accessed July 15, 2016. https://www.unocha.org/sites/unocha/files/2011%20OCHA%20Annual%20Report%20Final%20150dpi.pdf.

United Nations Framework Convention on Climate Change, The. 2015. "Updated Compilation of Information on Mitigation Benefits of Actions, Initiatives and Options to Enhance Mitigation Ambition." http://climateaction2020.unfccc.int/.

United Nations Framework Convention on Climate Change, The. 2015. "India and France Launch International Solar Energy Alliance at COP21." August 31, 2016. http://newsroom.unfccc.int/clean-energy/international-solar-energy-alliance-launched-at-cop21/.

United Nations Framework Convention on Climate Change, The. 2015. "Lima-Paris Action Agenda." August 31, 2016. http://newsroon.unfccc.int/lpaa.

United Nations Framework Convention on Climate Change, The. 2015. "Non-State Actor Zone for Climate Action." August 31, 2016. http://newsroom.unfccc.int/lpaa/nazca/.

United Nations Framework Convention on Climate Change, The. 2013. "Warsaw International Mechanism for Loss and Damage Associated with Climate Change Impacts." August 29, 2016. http://unfccc.int/adaptation/workstreams/loss_and_damage/items8134.php.

United Nations Framework Convention on Climate Change, The. 2012. "Report of the Conference of the Parties on Its Eighteenth Session, Held in Doha from 26 November to 8 December 2012." November 11, 2016. http://unfccc.int/resource/docs/2012/cop18/eng/08.pdf.

United Nations Framework Convention on Climate Change, The. 2011. "Conference of the Parties Seventeenth Session Durban, 28 November to 9 December 2011." November 11, 2016. http://unfccc.int/resource/docs/2011/cop17/eng/07.pdf.

United Nations Framework Convention on Climate Change, The. 2011. "Report of the Conference of the Parties on Its Sixteenth Session, Held in Cancun from 29 November to 10 December 2010. Accessed November 11, 2016. http://unfccc.int/resource/docs/2010/cop16/eng/07a01.pdf.

United Nations Framework Convention on Climate Change, The. 2009. "Report of the Conference of the Parties on Its Fifteenth Session, Held in Copenhagen from 7 to 19 December 2009." Accessed November 11, 2016. http://unfccc.int/resource/docs/2009/cop15/eng/11a01.pdf.

United Nations Framework Convention on Climate Change, The. 2008. "Bali Action Plan." Accessed August 23, 2016. http://unfccc.int/resource/docs/2007/cop13/eng/06a01.pdf.

United Nations Framework Convention on Climate Change, The. 2008. "Report of the Conference of the Parties on Its Fourteenth Session, Held in Poznan from 1 to 12 December 2008." Accessed November 11, 2016. http://unfccc/resource/docs/2008/cop14/eng/07.pdf.

United Nations Framework Convention on Climate Change, The. 2007. "Report of the Conference of the Parties on Its Thirteenth Session, Held in Bali from 3 to 15 December 2007." Accessed November 11, 2016. http://unfccc.int/resource/docs/2007/cop13/eng/06a01.pdf.

United Nations Framework Convention on Climate Change, The. 2006. "Report on Activities United Nations Framework Convention on Climate Change, Conference of the Parties 11 November 29 to December 6, Montreal, Canada 2005." Accessed November 11, 2016. http://www.arcticathabaskancouncil.com/aac/files/climate_change/UNFCCCCOP11.pdf.

United Nations Framework Convention on Climate Change, The. 2005. "Report of the Conference of the Parties on Its Tenth Session, Held at Buenos Aires from 6 to 18 December 2004." Accessed November 11, 2016. http://unfccc.int/resource/docs/cop10/10a01.pdf.

United Nations Framework Convention on Climate Change, The. 2004. "Report of the Conference of the Parties on Its Ninth Session Held in Milan from 1 to 12 December 2003." Accessed November 11, 2016. http://unfccc.int/resource/docs/cop9/06.pdf.

United Nations Framework Convention on Climate Change, The. 2003. "Report of the Conference of the Parties on Its Eighth Session, Held at New Delhi from 23 October to 1 November 2002." Accessed November 11, 2016. http://unfccc.int/resource/docs/cop8/07a01.pdf.

United Nations Framework Convention on Climate Change, The. 2002. "Report of the Conference of the Parties on Its Seventeenth Session, Held at Marrakesh from 29 October to 19 November 2001." Accessed November 11, 2016. http://unfccc.int/resource/docs/cop7/13a01.pdf.

United Nations Framework Convention on Climate Change, The. 2001. "Report of the Conference of the Parties on the First Part of Its Sixth Session, Held at the Hague from 13 to 25 November 2000." November 11, 2016. http://unfccc.int/resource/docs/cop6/05.pdf.

United Nations Framework Convention on Climate Change, The. 1999. "Report of the Conference of the Parties on Its Fifth Session, Held at Boon from 25 October to 5 November 1999." Accessed November 11, 2016. http://unfccc.int/resource/docs/convkp/kpeng.pdf.

United Nations Framework Convention on Climate Change, The. 1998. "Kyoto Protocol to the United Nations Framework Convention on Climate Change." Accessed October 29, 2016. http://unfccc.int/resource/docs/cop5/06a01.pdf.

United Nations Office for the Coordination of Human Affairs, The. 2016. "OCHA and Slow-Onset Emergencies." *OCHA Occasional policy Briefing series, No. 6*. Accessed September 24. http://www.reliefweb.int/sites/reliefweb.int/files/resources/report_36.pdf.

United Nations Treaty Collection, The. 2015. "The Paris Agreement, Paris, 12 December 2015." Accessed November 11, 2016. https://treaties.un.org/pages/ViewDetails.aspx?src=TREATY&mtdsg_no=XXVII-7-d&chapter=27&clang=_en.

United Nations World Commission on Environment and Development, The. 1987. "Report on the World Commission on Environment and Development: Our Common Future." Brundlandt Report. New York. Accessed November 16, 2016. http://www.un-documents.net/our-common-future.pdf.

Uitermark, Justus. 2005. "The Genesis and Evolution of Urban Policy: A Confrontation of Regulationist and Governmentality Approaches." *Political Geography* 24, no. 2: 137–163. doi:10.1016/j.polgeo.2004.07.009.

United States Department of Commerce, Economics, and Statistics Administration. 2013. "Economic Impact of Hurricane Sandy: Potential Economic Activity Lost and Gained in New Jersey and New York." *Office of Chief Economist*, https://rucore.libraries.rutgers.edu/rutgers-lib/43475/.

United States Department of Homeland Security. 2004. "Final Draft: National Response Plan." Accessed May 2, 2016. https://www.dhs.gov/xlibrary/assets/NRP_Brochure.pdf.

United States Department of Housing and Urban Development. 2008. "Indiana Income Limits." http://www.huduser.org/Datasets/IL/IL08/in_fy2008.pdf.

United States Department of State. 2016. "Major Economies Forum on Energy and Climate." Accessed August 23, 2016. https://2009-2017.state.gov/e/oes/climate/mem/index.htm.

United States Energy Information Administration. 2016. "Monthly Energy Review." Accessed August 2, 2017. http://www.eia.gov/totalenergy/data/monthly/pdf/sec1.pdf.

Vasudevan, Alexander. 2014. "The Autonomous City: Towards a Critical Geography of Occupation." *Progress in Human Geography* 39, no. 3 (April): 316–337. doi:10.1177/0309132514531470.

Vivekananda, Janani, and Neil Beat. 2017. "Coastal Megacities vs the Sea: Climate and Security in Urban Spaces." In *Epicenter of Climate and Security: The New Geostrategic Landscape of the Anthropocene,* edited by Caitlin E. Werrell and Francesco Femia. Washington, DC: The Center for Climate and Security. 2017. Accessed October 28, 2017. https://climateandsecurity.org/reports/.

Wakefield, Stephanie and Bruce Braun. 2014. "'Governing the Resilient City' Introduction, A New Apparatus: Technology, Government, and the Resilient city." *Environment and Planning D: Society and Space*. 32 no. 1 (February): 4–11.

Ward, Kevin. 2006. "'Policies in Motion', Urban Management, and State Structuring: The Trans-Local Expansion of Business Improvement Districts." *International Journal of Urban and Regional Research* 30, no. 1 (March): 54–75. doi:10.1111/j.1468-2427.2006.00643.x.

Watt, Paul. 2013. "'It Is Not for Us': Regeneration, the 2012 Olympics and the Gentrification of East London." *City* 17, no. 1 (February): 99–118. doi:10.1080/13604813.2012.754190.

Waugh, W. L. Jr. 2006. "The Political Costs of Katrina and Rita Disasters." *The Annals of the American Academy of Political and Social Science* 604, no. 1 (March): 10–25.

Wuebbles, D. J., D. W. Fahey, K. A. Hibbard, B. DeAngelo, S. Doherty, K. Hayhoe, R. Horton, J. P. Kossin, P. C. Taylor, A. M. Waple, and C. P. Weaver. 2017. "Executive Summary." In: *Climate Science Special Report: Fourth National Climate Assessment, Volume 1.* U.S. Global Research Program, Washington DC, USA: 12–34. doi:10.7930/JODJ5CTG.

Werrell, Caitlin, and Francesco Femia. 2017. *Puerto Rico and Climate Security Planning*. The Center for Climate Security: Exploring the Security Risks of Climate Change. Accessed September 29, 2017. https://climateandsecurity.org/2017/09/29/puerto-rico-and-climate-security-planning/.

Wood, George S., and Juan C. Judikis. 2002. *Conversations Community Theory*. West Lafayette, IN: Purdue University Press.

Yusoff, Kathryn. 2015. "Anthropogenesis: Origins and Endings in the Anthropocene." *Theory, Culture, and Society* 33, no. 2 (April): 1–26. doi: 10.1177/0263276415581021.

Zamora, Daniel. 2016. *Foucault and Neoliberalism*. Malden, MA: Polity Press.

Zimmermann, Karsten, Jasmin Boghrat, and Meike Weber. 2015. "The Epistemologies of Local Climate Change Policies in Germany." *Urban Research and Practice* 8 no. 3 (June): 303–318. doi:10.1080/17535069.2015.1051379.

Zomer, Alisa, and Angel Hsu. 2015. "Cities and Climate Change: Examining the Potential of Cities to Mitigate Global Climate Change." The Stanley Foundation and the Lyndon B. Johnson School of Public Affairs, University of Texas at Austin, *Key Regional Actors and Sector Opportunities for International Climate Change Cooperation.* Accessed June 25, 2017. http://www.stanleyfoundation.org/climatechange/Zomer-Yale-CitiesandClimateChange.pdf.

Index

About the Author

Dr. Julia Nevárez focuses her research on urban issues, urban development, public space, and technoculture. Her interdisciplinary analysis to urban living is evident in her contributive book chapters on topics including public space and globalization in Central Park; large video screens, public space and governmentality in Times Square; and public space and gentrification in Harlem. She is coeditor of: *On Global Grounds: Urban Change and Globalization* (with Gabriel Moser, 2009), which examines the emergent globalization processes of the early 2000s in different cities around the world. Her interest in theory as a way of conceptualizing contemporary urban space has focused on issues such as global urbanization, the networked society, critique of the global city, and social justice in the city. Her current work focuses on climate change preparedness and adaptation in urban areas. Her next book, *The Urban Library: Public Space and Urban Development*, will work on urban public space. Nevárez served as an Executive Board Member of the Urban Affairs Association. She is a professor and Sociology Coordinator at Kean University.

Lightning Source UK Ltd.
Milton Keynes UK
UKHW010403290519
343512UK00009B/270/P

9 781498 543750